Praise for

Essential Sustainable Home Design

It's all here. Everything you need to know about designing a sustainable home, for you, your family and the planet. Easy to read and understand while tackling a myriad of complex subjects. Well illustrated and filled with useful information.

—Martin Hammer, architect, co-author of the Strawbale, Light Straw-Clay and Tiny Houses appendices in the *International Residential Code*

Chris has given us a wonderful gift in this book. The rational approach to design criteria is excellent and thorough, measured with humour and honesty based on his experience building sustainably over the past 20 plus years. I will be referring prospective clients and colleagues alike to *Essential Sustainable Home Design*.

—Tim Krahn, LEED AP, P. Eng.

Once again, Chris Magwood has put his deep and broad experience to excellent use in creating this book. The book is both comprehensive and concise, covering in a highly useful format, all the important aspects of the process of seeking sustainable goals in designing a house. Chris couples his extensive and rigorous approach to research, experimentation, and learning, with his gift for teaching and communicating what he's learned. The result is another terrific addition to the sustainable design and building literature.

—David Eisenberg, Director, Development Center for Appropriate Technology

Finally! An up-to-date overview of the most important approaches and questions to address when making a more sustainable home. This book is essential in navigating the overwhelming field of sustainable house design and distills it down to its basics, allowing one to make informed decisions about creating their home.

—Ben Falk, MALD, author, *The Resilient Farm and Homestead* and *Whole Systems Design*, LLC

I found this to be a useful and comprehensive guide to the entire process of creating a sustainable and resilient home by a leading exponent of sustainable building practices. The author correctly stresses the importance of process — determining the objectives of the project up front, specifying clear project goals, and putting together an appropriate team to realise these goals. Also stressed are the decision-making processes and the importance of understanding the big picture implications of decisions. These are essential but often overlooked aspects of creating a sustainable home.

—Dr. Mark Gorgolewski, Chair of the Department of Architectural Science, Ryerson University, Toronto.

This wonderful book by Chris Magwood offers a wealth of knowledge about the materials, systems and processes needed to design better homes. Two things set this book apart from other "green" guides: The Approach section gives you the tools you need to create your own set of goals and criteria, and the section on Materials and Systems offers a comprehensive and unbiased review of both "green" and mainstream materials.

—Larry Strain, FAIA LEED AP, SIEGEL & STRAIN Architects

Creating a home from scratch is a rare opportunity to express our deepest values. If those values include health and planetary ecology the process can be especially daunting because these are not the primary values of the conventional housing market, and the "Green" building world is multifaceted with information that often seems contradictory. In *Essential Sustainable Home Design*, Chris has written an intelligent synthesis of all things "green", presented in an easy-to-understand format to help future home owners make the best choices for every stage in creating their new home.

—Paula Baker-Laporte FAIA co-author,
Prescriptions for a Healthy House and *The EcoNest Home*

essential
SUSTAINABLE HOME DESIGN

essential SUSTAINABLE HOME DESIGN

a complete guide to goals, options, and the design process

Chris Magwood

Cover design by Diane McIntosh.
House plans: Cindy McCaugherty of RainCoast Homes Drafting and Design.
Section 1 illustrations by Dale Brownson. Section 2 illustrations by Chris Magwood.
Background photo AdobeStock_95944284.

Printed in Canada. First printing July 2017.

Any other inquiries can be directed by mail to:
New Society Publishers
P.O. Box 189, Gabriola Island, BC V0R 1X0, Canada
(250) 247-9737

LIBRARY AND ARCHIVES CANADA CATALOGUING IN PUBLICATION

Magwood, Chris, author
Essential sustainable home design : a complete guide to goals, options, and the design process / Chris Magwood.

(Sustainable building essentials)
Includes bibliographical references and index.
Issued in print and electronic formats.
ISBN 978-0-86571-850-0 (softcover).—ISBN 978-1-55092-645-3 (PDF).—ISBN 978-1-77142-240-6 (EPUB)

1. Ecological houses—Design and construction. 2. Sustainable design. 3. Dwellings—Environmental engineering. I. Title. II. Series: Sustainable building essentials

TH4860.M27 2017
690'.8047

C2017-904262-9

C2017-904263-7

Funded by the Government of Canada | Financé par le gouvernement du Canada | Canada

New Society Publishers' mission is to publish books that contribute in fundamental ways to building an ecologically sustainable and just society, and to do so with the least possible impact on the environment, in a manner that models this vision.

New Society Sustainable Building Essentials Series

Series editors
Chris Magwood and Jen Feigin

Title list

Essential Hempcrete Construction, Chris Magwood
Essential Prefab Straw Bale Construction, Chris Magwood
Essential Building Science, Jacob Deva Racusin
Essential Light Straw Clay Construction, Lydia Doleman

See www.newsociety.com/SBES for a complete list of new and forthcoming series titles.

THE SUSTAINABLE BUILDING ESSENTIALS SERIES covers the full range of natural and green building techniques with a focus on sustainable materials and methods and code compliance. Firmly rooted in sound building science and drawing on decades of experience, these large-format, highly illustrated manuals deliver comprehensive, practical guidance from leading experts using a well-organized step-by-step approach. Whether your interest is foundations, walls, insulation, mechanical systems, or final finishes, these unique books present the essential information on each topic including:

- Material specifications, testing, and building code references
- Plan drawings for all common applications
- Tool lists and complete installation instructions
- Finishing, maintenance, and renovation techniques
- Budgeting and labor estimates
- Additional resources

Written by the world's leading sustainable builders, designers, and engineers, these succinct, user-friendly handbooks are indispensable tools for any project where accurate and reliable information is key to success. GET THE ESSENTIALS!

Contents

Acknowledgments

I AM EXTREMELY LUCKY AND GRATEFUL that the "rock stars" of the natural building world are simultaneously true superstars of forward thinking about the care of people and the planet and humble, approachable, brilliant, and most often hilarious people who have built an extended family that I value greatly.

I am equally blessed to have a supportive and caring and patient and engaged family. Together, they have made me who I am, and there aren't really thanks enough for that.

For their help with this manuscript, I'd like to thank Jen Feigin, Jacob Deva Racusin, Deirdre McGahern, Erin McLaine, and Shane MacInnes.

I start out on this road,
I only know what is not here.

I am a universe in a handful of dirt,
whole when totally demolished.

Talk about choices does not apply to me.
While intelligence considers options,
I am somewhere lost in the wind.

— Rumi

Chapter 1

Introduction

I BEGAN MY 20-YEAR FORAY into the world of sustainable building as an idealistic and completely inexperienced amateur wanna-be homeowner — just a guy with big dreams, very limited means, and almost no idea of the complexity of the task I was attempting. Armed with one book (and no internet!) and great intentions and expectations, I plunged my family into a long-term adventure that changed all of our lives.

Though I wouldn't trade my personal experience for any other, I will gladly attest to the flaws of my naive approach. In fact, so flawed was this way of doing things that for two decades I have centered my life around helping others find their way to their own dream green home without hitting as many of the snags and making as many mistakes as I did.

There are many times in our lives when we make rash decisions and don't adequately prepare ourselves for a task. Most of the time, making a less-than-ideal choice isn't that big a deal — a poor choice can get chalked up to "live and learn." However, poor choices in home building can be extremely costly, and the results can have real and serious implications for decades to come. When your life's savings and a vast amount of your time and effort are on the line (not to mention large quantities of the planet's current and future resources), "oops" is not a word you want to hear! The world of homebuilding is not a place you want to wander in blind and be directed by hard knocks.

Sadly, I have watched an awful lot of people plunge into homebuilding only to make the same, predictable, costly and demoralizing mistakes that I did. I have seen many homes built well-over budget that also underperformed — never meeting the high expectations their owners had at the outset.

As much as I'd like to be able to offer a "silver bullet solution" that would guarantee quick-and-easy results, I'm afraid there is no fast track, sure-fire method to figuring out how to build yourself the best possible home. I have spent two decades deepening my knowledge of how to make a really good building, and I still have lots to learn. It's a vast subject, and the determining factors are many: climate and site, local regulations, available resources and skills, and, of course, budgets, which vary widely, as do individual considerations of comfort and aesthetics. There is no one "perfect" way to balance all of these factors; each project requires unique adaptations.

The uniqueness and "specificness" of homes — which I believe is essential for making a house into a home — has been largely abandoned for a one-size-fits-all simplicity that suits the needs of the large-scale construction industry but has done a large disservice to humans, the built environment, and the planet. This is not to say that good homes cannot be simple, but rather that the pathways to arrive at a good home design are as varied and many as the number of people who need and want homes.

Challenges and Rewards

We are at the beginning of a major period of disruption in the building industry. Pressures from many directions are forcing important changes in the practice of home design and construction, including more stringent energy efficiency codes, concerns about occupant health, and the imperative to reduce carbon footprints and change to clean energy sources — all of which are having dramatic effects on how we build. The cost of property, materials, and labor has been on a steep incline for over a decade. It is a constant challenge to find the best ways to meet a reasonable budget target and achieve a high level of performance, and if the intention is to also use the healthiest possible materials, the challenge is amplified. Add in a dash of aesthetics and a fair share of bureaucratic red tape and regulatory hurdles, and you have a serious challenge on your hands.

The effort required to prepare yourself for this particular adventure is great, and the decision to move forward must begin with acknowledging that you are about to engage in a process that will be all-consuming for at least a couple of years. If this idea doesn't appeal to you, don't go down this path. Give this decision the weight it deserves. Put it on par with decisions of the magnitude of changing your career, going back to school, or moving to a new city or country.

But before I discourage you from even considering setting foot on this path, I should mention the incomparable satisfaction that comes from settling down for an evening in a home that you have designed and built for yourself, your family, and your friends. In a world where many of the archetypal "coming of age" moments are absent or watered down, weathering your first literal storm sheltered in your own home is a great and deep satisfaction. And if you can manage to get through that storm with the lightest possible footprint on the planet and the healthiest possible environment surrounding you, the satisfaction goes beyond just a personal achievement and becomes something that will be an integral part of your life and a legacy that will live long beyond your time on this planet.

It's my hope that you can make your project a forward-looking legacy, one that provides you and your family with all that they need while also achieving the goal of sustainable development — nicely defined by the UN's World Commission on Environment and Development: "Sustainable development is development that meets the needs of the present without compromising the ability of future generations to meet their own needs."[1]

Section 1

Developing a New Approach

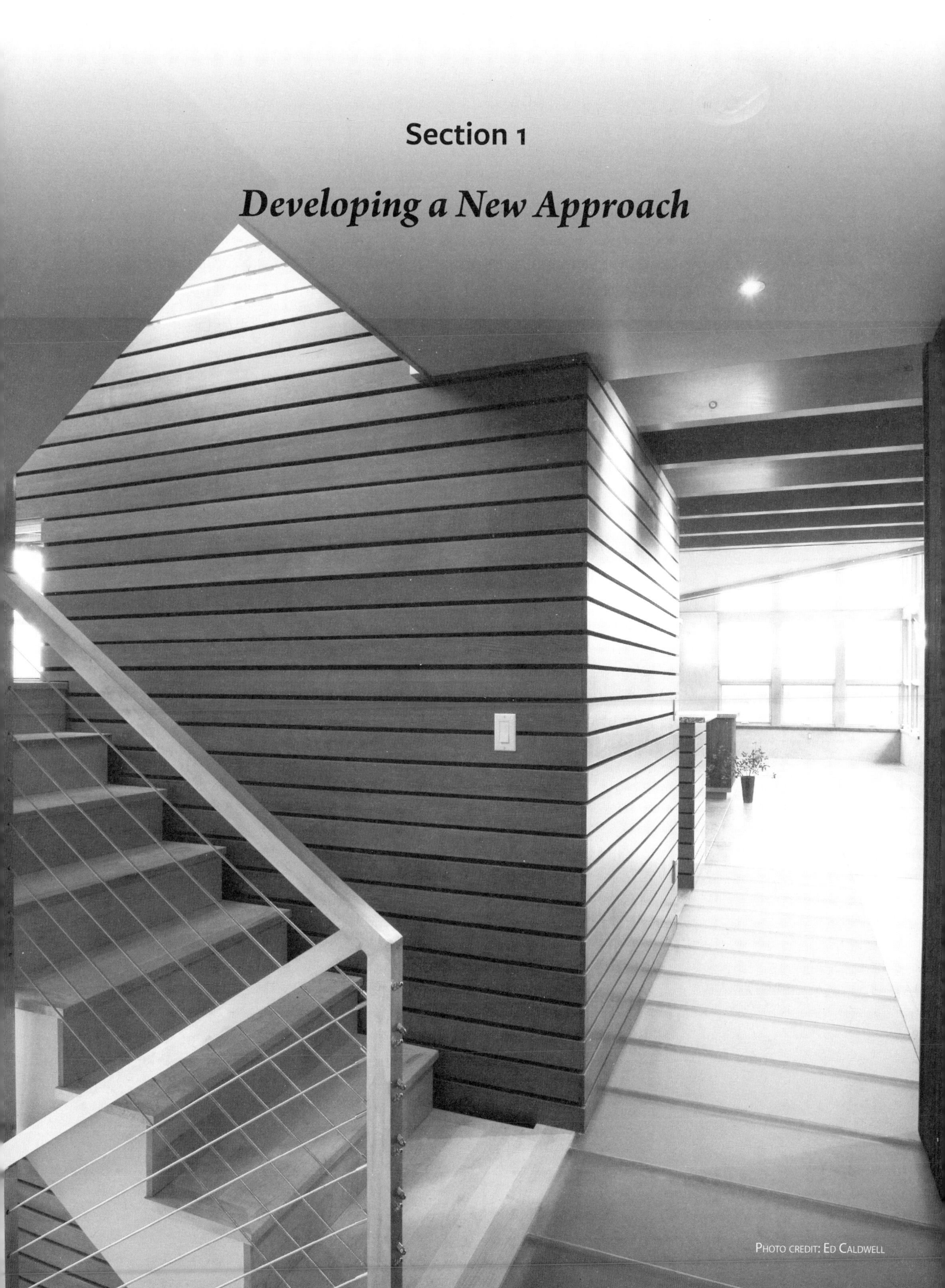

Photo credit: Ed Caldwell

Chapter 2

Decision-Making

MOST BOOKS ABOUT HOME DESIGN jump right into the process of actually putting a house design to paper (or, in the modern context, to a software program). We're not going to get to that stage for a while. Of course, this is an essential part of designing a sustainable home, and it will be covered here in reasonable detail, but designing a sustainable home requires much more than putting lines onto a drafting board in the right order. It needs to start not with *drawings,* but with *goal setting.*

The fact that you are reading a book about sustainable home design indicates that you have an interest in setting a goal for your project that in some way addresses key issues of personal and/or societal sustainability. However, each of us probably means something different when we use such a term — and the words we use may themselves be different. Terms such as *sustainable, healthy, eco-friendly, natural, green, environmentally-sensitive,* and *net zero* are often used — sometimes interchangeably — to describe the kind of better home an owner desires.

Measurable outcomes as the basis of decision-making

It just won't work to begin your home design based on a couple of turns of phrase and a vague notion of what they mean to you. This book is not a treatise on semantics, so we're not going to try to define any terminology for you. Instead, we are going to focus on defining the actual goals that you are setting out to achieve when you use such terminology. Rather than relying on simple taglines, we're going to focus on well-defined goals and measurable outcomes. If you can knowledgeably set the targets you'd like your project to hit, your chances of succeeding are vastly increased.

Understanding Your Objectives

The unique intention of this book is to help you understand, define, and refine your objectives so that you can make informed choices — from initial siting considerations, through personnel decisions, and down to the level of individual material and system selections. It is critically important to acknowledge that your goals can be undermined by poorly informed choices at any stage of your project. If you fail to set appropriate goals at the "meta" level, then the chances of your project succeeding are greatly reduced, and if you do not ensure that each individual choice you make — throughout the process — adheres to the goals you've set, the results will likewise be undermined.

For example, it is the goal of many homeowners to create an energy-efficient home. At the

A good building means different things to different people.

"meta" level, this would involve setting a hard target — for example, meeting current Energy Star or Passive House standards. With this kind of clearly definable goal, it becomes possible to make informed and appropriate design, material, system, and personnel decisions. In this case, the clear goal of meeting a particular efficiency standard would narrow down the selection of design professionals to those trained to meet that standard, and it would help with the selection of insulation materials, doors and windows, and mechanical systems appropriate to meeting the standard. Any choice that would subvert the larger goal is easily discarded in favor of one that harmonizes with the overall intention of the project.

Much of this book is dedicated to helping you set these overarching goals because clarity on the relatively small list of guiding issues greatly simplifies the very large list of decisions that need to be made during a design/build project.

Learning to think "sustainably"

What makes it onto the scales during decision-making, and why?

The mechanics of decision-making are quite straightforward: we weigh up all the pros and cons of competing choices, and we choose the option that has more pros than cons.

The process is no different when you are aiming to make a sustainable home, but the way in which we draw up the list of pros and cons includes some factors that are very often overlooked. We have to check some deeply ingrained biases if we are going to make the best choices.

New ideas versus established solutions

When considering new solutions, our approach tends to be one of two extremes. There are those of us who are inclined to accept the promises of a new solution without applying much rigor toward finding out if the promises are true, and there are those of us who tend to dismiss the promises of new solutions, also without the application of much rigor in our examination.

One of the first questions I am often asked when discussing new approaches to building is: "Does [insert name of material or system] *really* work?" This is an entirely appropriate and important question to ask, and it's not unusual that we should have this question when faced with something new. The "new solution paradox" is that we typically *only* ask it of new solutions, and completely fail to question existing, accepted solutions. There is an assumption that the ideas, materials, and systems we use commonly have somehow been "proven" to work, that they have been rationally measured and found to be the best means to achieve a particular end. In the realm of building materials and systems, however, development, testing, and establishment of industry standards have been far from rational and well-proven processes. As most readers of this book are already aware, most of our accepted solutions have not been developed with any coherent ecological principles or human health ideals in mind.

We tend to expect *new* ideas or technologies to live up to *unrealistically high* standards, while at the same time *normalizing* existing ideas or

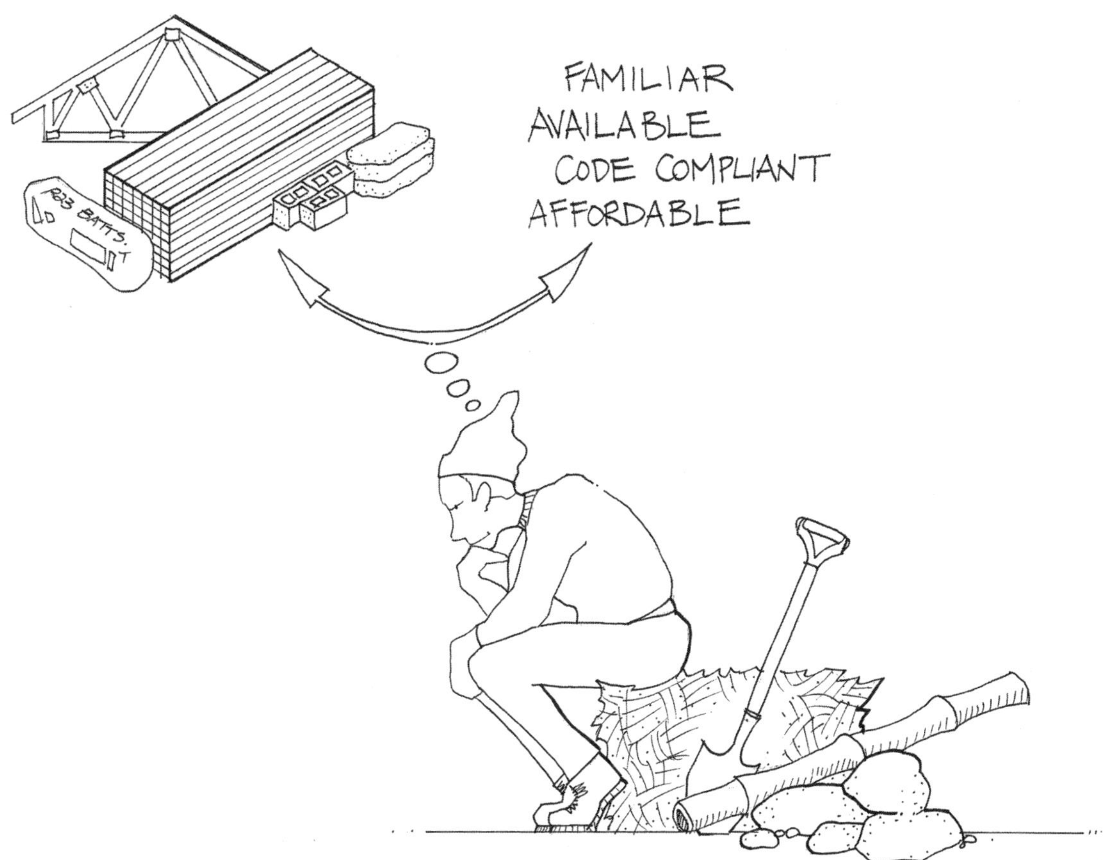

Familiar expectations often match with familiar criteria.

technologies that are inherently, deeply *flawed.* If we're interested in making improvements in our buildings, it is critical that we hold both "accepted solutions" and "alternative solutions" to the same standards, using the same criteria before coming to conclusions. It is great to try to be objective about the choices we make, but it is essential that we apply the same objectivity to *all* our choices, including those that are so normalized that we don't see them as choices, but as inevitabilities.

Assessing inevitable flaws

There is no such thing as an idea or technology with no flaws. Recognizing this simple point is key to being able to consider new ideas fairly and thoroughly.

To prepare our minds for considering new building material ideas, it is helpful to think about one of the most trusted materials in the North American construction industry: wood. We rely on structural wood framing for a huge percentage of our residential and commercial buildings, and we use wood for finishes on the interiors and exteriors as well. Yet this trust in wood comes from the fact that its use is "normal" for us. If we were to try to introduce wood as a brand new building material today, it would face an uphill battle. Skeptics would raise all kinds of issues, pointing out that wood:

- Burns easily
- Rots naturally when exposed to moisture
- Is eaten by a wide range of common insects
- Is a great growing medium for mold
- Expands and contracts considerably depending on moisture content
- Twists and splits when drying

- Has strengths that vary widely depending on species and growing conditions
- Has strengths that vary widely depending on milling, drying, and storing processes
- Is often grown far from where it's used

Yet wood has come to serve us very well as a building material; along with its flaws, it also has a wide range of great properties that encouraged us to work to overcome all the flaws. We've now normalized it and built an entire successful industry around a highly flawed material. And though building codes and the lumber industry can provide plenty of data to justify the use of wood for its good points and minimize its flaws, this "proof" of the validity of wood as a building material came long after it had been widely adopted. As with so much of what is "normal" to us today, adoption was based on need, convenience, and field testing, not rational analysis.

Focusing on potential flaws can destabilize a project.

When we now attempt to introduce a new material that has even a small number of the very same flaws inherent in wood, we find ourselves up against naysayers who allow the existence of flaws to blind them to the strengths of the new material and the possibilities for being able to overcome any flaws.

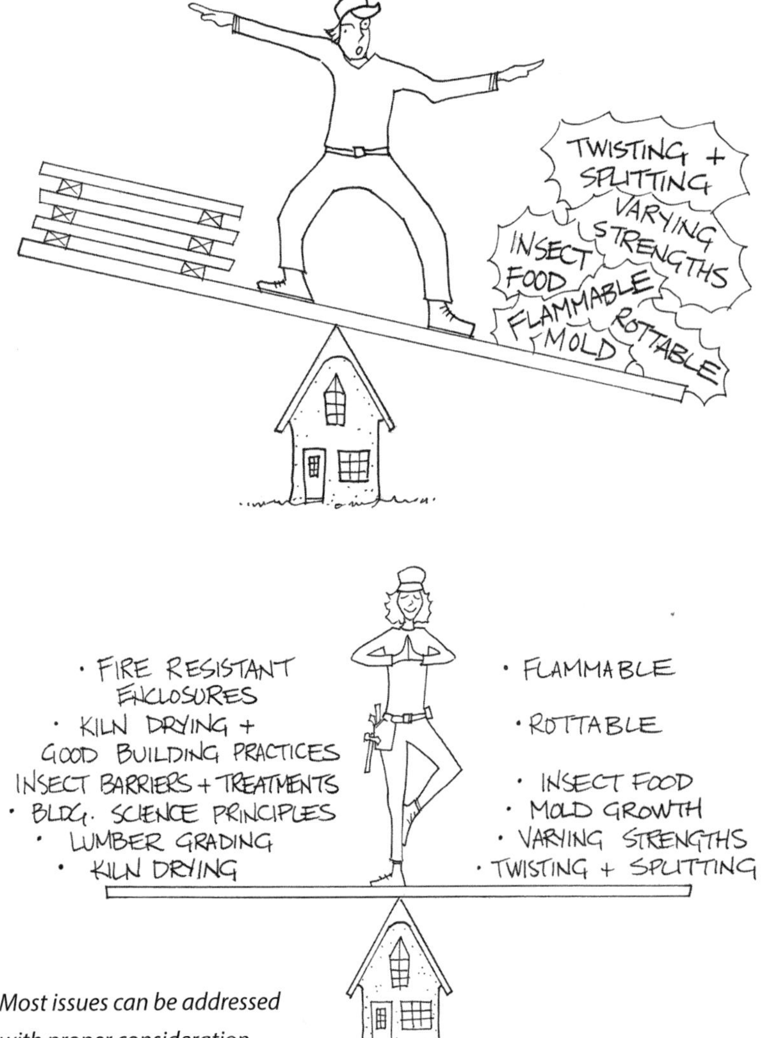

Most issues can be addressed with proper consideration.

Micro vs macro views

Once we recognize that there are no options without some inherent flaws, we need to be able to see these building problems and solutions at two different levels — the *micro* and the *macro*. The vast majority of building-related decisions are viewed at the micro level, and involve assessing choices between competing materials and systems (often in the form of products).

Let's look at one example of how the difference between accepted and alternative solutions and micro and macro perspectives can play out when making building choices: *flush toilets versus composting toilets*. Many homeowners considering a more sustainable home will consider this issue at some point in their decision-making process.

Let's first ask if both solutions "work."

We tend to assume without much questioning that the flush toilet option works. After all, every building in the past half-century has used some version of this technology; we use them every day ... so *of course it works*. However, it is unlikely that we have gone through life without experiencing at least one unpleasant backing-up and overflow experience with a flush toilet. In fact, it's likely to have happened several times to each of us, and it's been an unpleasant situation to deal with, not mention extremely unhygienic. But even such dramatic failures don't lead us to consider that flush toilets *don't work.*

On the other hand, most of us have little or no experience with composting toilets, and our judgment of whether or not this entire range of options "works" will be based on either limited personal knowledge or on the received opinions of others. Proof that composting toilets "don't work" are mostly based on reports of unpleasant odor and a general revulsion at dealing with human waste. These "failures" lead us to conclude that composting toilets don't work.

Of course, we understand there are reasons a conventional flush toilet gets clogged, and we know it can be remedied. It is a malfunction. But it must be remembered that it is a malfunction that is almost 100% likely to occur. And should it be recurring, we will blame the model of toilet (they're not all created equal) or a systemic problem with the plumbing attached to the toilet, but not the entire notion of flush toilets.

The important thing is to remember that the same is true of the composting toilet — systems that have negative issues are experiencing a malfunction that can be remedied. And systems that experience recurring problems are indicative of a faulty model design or a systemic problem with the use of the toilet. But a few issues with a few composting toilets don't negate the fact that the majority of composting toilets work perfectly well most of the time — just like flush toilets.

So, at this point, we can identify "micro" level flaws with both systems that can produce unpleasant encounters with human excrement. Next on the comparison list is likely to be cost. Here, it would seem that the flush toilet is the clear winner, as effective composting toilet systems appear to be many times more expensive. But before the flush toilet is declared the

It is possible — and even likely — for both flush toilets and composting toilets to malfunction at some point. One doesn't work better than the other, but they have different means and reasons for malfunctioning.

clear winner on this point, we have to look at the complete system costs for both. While a flush toilet unit is not very expensive, a full cost accounting would need to look at the costs for a septic system (in rural areas) and sewage connection fees and ongoing water/sewage charges in urban areas. The elimination or drastic reduction of those costs will likely put the composting toilet system in a similar cost bracket, especially if long-term costs are factored in.

Full-system costing must be used to create fair comparisons for budgeting.

Now comes the important step for the sustainable builder. We must move our focus from micro concerns like functionality, form, and cost and look at the larger ecological implications of our choices. Often, these implications are not immediately obvious and rarely used as a point of comparison, but it is crucial to include them if our intent is to make choices that are better for ourselves, our children, and the planet.

The flush toilet, viewed through this lens, is an ecological disaster. For private septic systems, the numbers are discouraging. "According to the US EPA, failure rates for on-lot sewage systems across the country are reported at 10 percent annually."[2] This means that millions of systems are leaching noxious quantities of contaminated effluent into surface- and groundwater supplies. These numbers indicate that within ten years, almost all systems will experience failure, and failure results in the pollution of our collective water supply. The numbers aren't any better for municipal sewage systems. The Sierra Legal Defense Fund's 1999 Sewage Report Card states: "Over one trillion liters of primary or untreated sewage is collectively dumped into

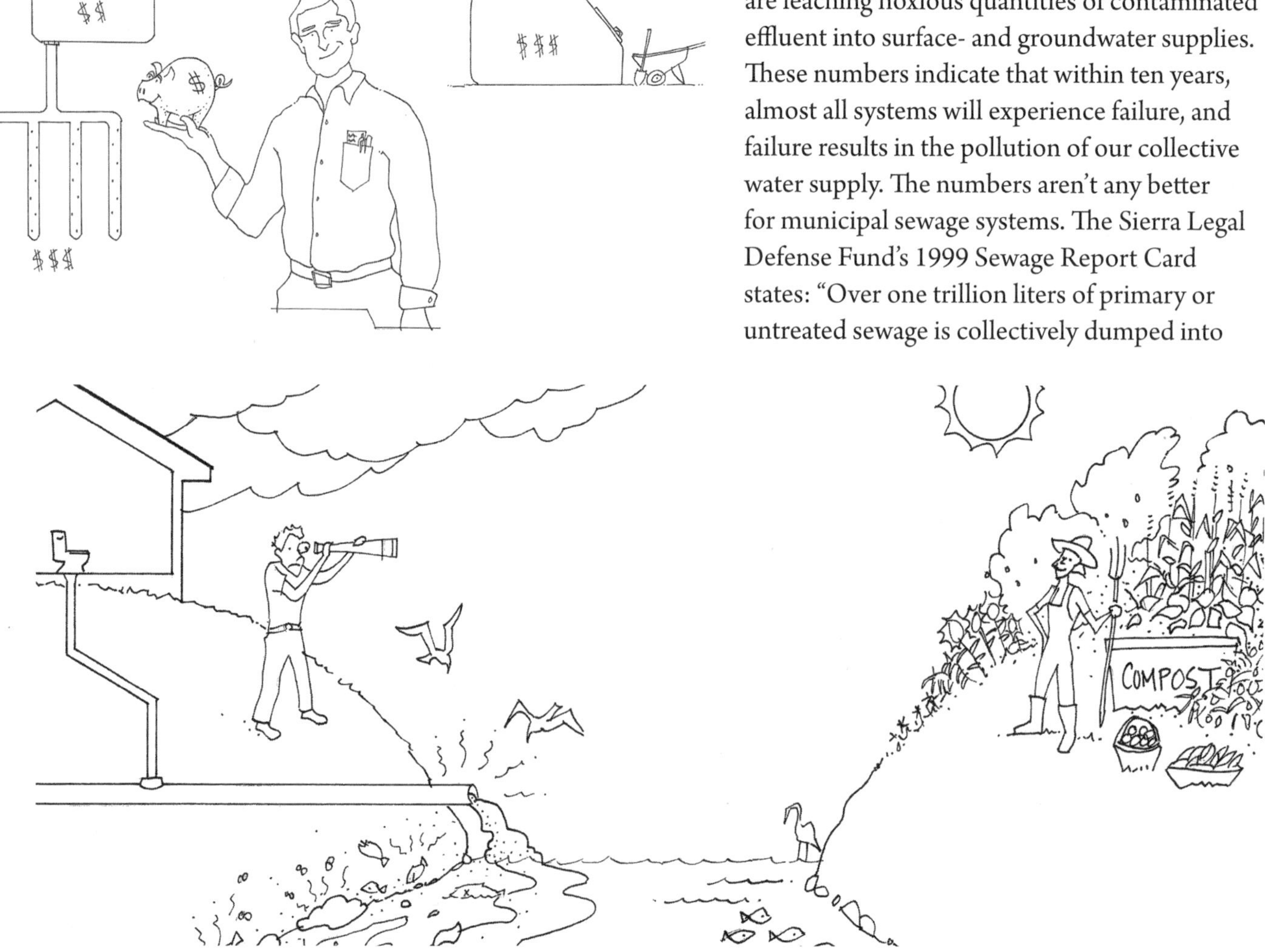

The macro comparison of flush toilets versus composting toilets shows a vastly different impact on human health and the planet. If these issues are important to us, then there is a clear winner in this comparison.

our [Canadian] waters every year by the [21] cities evaluated in this report. This volume would cover the entire 7800 kilometer length of the Trans-Canada Highway to a depth of nearly 20 metres — six stories high."[3] With ten times the population, US sewage dwarfs this volume. In addition, trillions of liters of treated water are used to flush all our toilets and create all this sewage, and this too has significant ecological and financial effects. Clearly, this accepted solution is not much of a solution when viewed on the macro level.

Composting toilets, when working properly, make compost that can be used to amend soil. A search of the database of The Center for Disease Control in the United States does not turn up any evidence of a single human illness attributed to composting toilets. The Center undertook a study of composting toilets at a national park in Arizona, and found that workers who clean and maintain the composting toilets were "more at risk for extreme heat, bee and scorpion stings, spider bites, and hantavirus from rodent nests and feces" than from the composted waste they were dealing with.[4] At the very least, composting toilets present a lower risk to the environment than the sewage created by flush toilets, and at best they have a positive impact in the rebuilding of soils.

Technologies develop the longer they are in use. Some sustainable building technologies are in the "Model T phase"—functional but not developed and refined. There is no reason to think that they will not reach the "self-driving phase" as they become more popular.

Being an early adopter

Widening our outlook to include macro considerations like environmental impact will probably lead us to choices that are not yet mainstream options. It is important to remember that new ideas and technologies do not leap out of the gate fully formed and perfectly developed. It is wise to consider where a particular option may be situated in terms of its development arc, and to understand where we may or may not want to intersect with that arc. There are definitely rewards for diving in and being among the earliest of adopters of a particular material or system, but there are also drawbacks. Early in the development phase expect higher costs, less availability, and the need to undertake some troubleshooting to make things work right. As the idea develops, cost and availability tend to improve, as do ease of installation, functionality, and maintenance requirements.

The "Yah, but..." response

Early in the development phase of new ideas and systems, there are many naysayers. The "*Yah, but…*" response is a common means of acknowledging that a new idea has some merit, but dismissing it based on its current state of development. To illustrate, we can continue using the composting toilet versus flush toilet comparison. It would be easy to dismiss the notion of widespread use of composting toilets as being an idea with positive possibilities, but impractical to consider as a mainstream option.

This outlook ignores the trajectory of all developing ideas, which start as good ideas and then move into prototype and testing phases, followed by small-scale market penetration and, in some cases, widespread adoption. Composting toilets may have limited appeal because of inherent revulsion about dealing with human waste. However, it is possible to imagine the feasibility of service contracts for removing composting toilet contents and handling them at a central facility in much the same way that we currently handle solid waste and household recycling. Commitment to a better idea is what spurs the development of practical solutions — the notion of recycling came before the invention of coordinated neighborhood pickups in specialized trucks.

The "Yah, but" outlook also ignores the ways in which social mores can influence development of ideas. In many parts of North America, we applied laws and a high degree of social pressure to induce dog owners to pick up their animals' feces from lawns, parks, and sidewalks. In a matter of a decade, this social pressure resulted in a new norm, one in which dog owners get "up close and personal" with their animals' waste (in a way that is much more visceral and off-putting than dealing with a composting toilet!). If we seriously undertook a social plan to make it inexcusable to foul our waterways with human waste, we could achieve the same type of result.

If it's possible to convince people to do this, it's possible to get people to deal with clean and simple humanure composting.

Doing things better not the same as doing better things

Throughout history we have collectively refined ideas and technologies, even those with significant hurdles (such as the automobile, with the vast infrastructure required to support its use and its devastating toll on humans and the environment). We're good at improving the micro-considerations, but generally terrible at intentionally setting macro-goals.

In the realm of building materials, cement followed the standard development arc. Early cement products were extremely labor- and fuel-intensive and far from reliable, but the benefits of having a quick-setting material that is easily formed and potentially strong encouraged us to work through all kinds of issues to arrive at this moment; the modern cement industry now offers well-formulated products that are widely available and cost-effective. Many of the "alternative solutions" presented later in this book are at the beginning of that same development curve; this doesn't make them impractical or impossible to implement, but it does require an acknowledgement that there is improvement to come in the future. The day may not be that far away when a builder can order a highly specialized clay plaster or clay floor mix from a local batching plant and have it delivered and placed with the same degree of mechanization as concrete! This is already true in Japan; it could just as easily happen here.

Personalize the solution

Looking at building solutions through a wider, ecological perspective can radically change how we consider our options on an individual basis. But each homeowner and builder will have their own unique outlook and therefore make different choices. In the next section of the book, we will look at developing a personalized list of specific criteria, and attempting to keep the widest possible perspective on the impacts our choices might have on ourselves, our children, and generations to come.

There is no reason that ecologically friendly, clay-based building materials can't be produced, delivered, and used with the same efficiency enjoyed by concrete materials.

Chapter 3

Defining Goals and Criteria

"I WANT A BEAUTIFUL, AFFORDABLE, COMFORTABLE, HEALTHY GREEN HOME." On this kind of premise, thousands of people set out to make their dream home a reality. In many cases, the prospective homeowners have looked around and found the conventional offerings to be lacking, and decided to forge ahead and do it themselves. Though the desire for such a home is a basic and simple one, the execution is most definitely not. If we are going to make this dream come true, we must first begin by understanding exactly what it is we are trying to achieve.

A lot of us find it easiest to define our sustainable housing ideas in the negative, as a reaction against the type of conventional housing we commonly see. It is easy to demonize the modern construction industry. From our current vantage point, the homes it produces have many measurably negative effects on the environment, on occupants, and on society. But really, the industry has spent a half-century or so doing exactly what regulators and the buying public has asked them to do. The industry didn't develop with the express intention of spreading ecological and social ruin and destruction — though in the pursuit of the collective goals we have set for our housing needs, much has been ruined and destroyed.

In general, we have asked for and received homes that meet the following criteria:

- Compliance with minimum code standards
- Cost that the market will bear (and therefore material and labor inputs that match cost expectations)
- Consistent aesthetics that the market demands/expects
- Level of energy efficiency that regulators enforce or the market demands
- Valuation that lenders and insurers demand (durability and maintenance based on typical mortgage cycle)
- Design and construction performed by professionals, not homeowners
- As little maintenance and homeowner engagement as possible
- Location and siting that match expectations and regulations

If these are, indeed, the goals set out for the construction industry, then the industry is doing an excellent job of meeting them, while ensuring a profit margin for developers and contractors adequate to keep the whole enterprise rolling along.

What do we want?

If we want something different, we have to identify different criteria, and be able to define and explain our goals so they can be successfully achieved. If we are seeking houses that are "better," we are typically interested in broadening our criteria — using the wider vision discussed in the previous chapter — to include these additional elements:

- Ecosystem impacts that are minimized or eliminated
- Embodied carbon and energy footprint that is minimized or eliminated
- Energy efficiency that exceeds code minimums, with reduced or eliminated need for fossil fuel input
- Indoor air quality that exceeds code minimums, with reduced or eliminated sources of toxins and allergens

- Locally sourced materials and labor, with reduced or eliminated need for exploitive labor practices and excessive transportation needs
- Waste reduction or elimination during construction and over the building's lifespan
- Resilience in the face of climate change
- Aesthetics that are personally desirable

These two lists of conventional and "alternative" criteria are not mutually exclusive — the issues currently addressed by the construction industry are important to all builders. In order to create a new, more sustainable home, or to significantly renovate an existing home, we will need to combine these lists into a comprehensive whole. Experience has shown that most green building projects address the following list of concerns, with greater or lesser emphasis on each category:

1. Ecosystem impacts
2. Carbon and energy impacts
3. Energy efficiency
4. Indoor environment quality
5. Waste
6. Resilience
7. Occupant input and durability
8. Material costs
9. Labor costs and sources of labor
10. Code compliance
11. Aesthetics

The path to a successful and satisfying building project requires us to fully understand our goals. That journey begins with a comprehensive understanding of each of these goals. Once we understand our goals, we can then define exactly how we envision meeting them.

The Concerns

1) Ecosystem Impacts

For many people, the notion of a green home begins with reducing the impacts the home will have on the planet's ecosystem. This sounds simple, but a building can require hundreds of different materials that come from sources all over the world, each with unique impacts on the ecosystem. Often, there are two distinct phases to investigate: the harvesting of raw materials, and the manufacturing process(es).

How can we assess our impacts? Important questions to ask about each material include:

Is the material based on renewable or nonrenewable resources?

- *Renewable resources* are those that can be regenerated through biological processes.
- *Rapidly renewable resources* — Like straw, hemp, cotton, and mycelium — are regenerated within a single growing season. Other renewable resources — like trees — can take decades to regenerate.
- *Nonrenewable resources* are those that exist in finite quantities on the planet, and for which there is no ability to generate more of the resource biologically.
- *Abundant nonrenewable resources* — Like limestone and clay — exist in large quantities and may be relatively straightforward to access.
- *Scarce nonrenewable resources* — Like precious metals — exist in relatively small quantities and can be difficult to locate and access.

Are the renewable resources actually being regenerated?

Just because a material *can* be renewed, doesn't mean that it *is* being renewed. Investigate sources to determine if replanting/replenishing is actually being done.

Are the renewable resources being regenerated responsibly?

- Just because a material is being renewed doesn't mean it is being done in an appropriate manner. Monocrop agriculture and single-species tree farms are typically less desirable ecosystems than organic agriculture and diverse forest replantation.

What is the extraction process for nonrenewable resources?

- The majority of nonrenewable resources are mined from the earth in some fashion, but the processes are invasive and destructive to different degrees. Even the same resource may be extracted with varying impacts in different places, so be sure to research the specific source of your material.
- *Fossil fuels* — Used to make plastic building materials — exist in relatively large quantities, but are being depleted quickly. The extraction of oil has a long list of negative ecosystem impacts.

While it may be tempting to value renewable resources over nonrenewable, the ecosystem impacts must be considered carefully for each material, as the overall ecosystems impacts may be higher for some renewable resources. It is most likely that a home will require the use of both types of resources, so both categories will need consideration.

What ecosystem disruption does the harvesting of this material cause?

- Are previously untouched ecosystems being destroyed or significantly altered?
- Are existing habitats of animal species being displaced?
- Are waterways being diverted or interrupted?
- Is the landscape being altered in significant ways?
- Is biodiversity being compromised?
- What effluent, discharge, or runoff is created? In what quantities and of what toxicity?
- What solid waste is created? How is it dealt with?

What types of pollution are associated with the manufacturing process?

- Is airborne pollution released? If so, in what quantities and of what toxicity?
- Is waterborne pollution released? If so, in what quantities and of what toxicity?
- Is pollution distributed or buried in soils? If so, in what quantities and of what toxicity?
- Are toxic conditions created for workers?
- What kinds of fuels are used to power the manufacturing process, and what types of pollution are associated with harvesting and using those fuel sources?

It is important to consider these types of pollution broadly and inclusively. For example, the use of any type of plastic or foam material inherently includes the pollution that is caused by oil exploration, extraction, spills from pipelines and wells, and transportation accidents, as well as the types of effluent emitted during all phases of processing, not just the final manufacturing process. Similarly, the impacts of agricultural products like straw and hemp need to include the full production and in-situ effects of fertilizers, pesticides, and herbicides used in the growing cycle.

What kind of water use and treatment are associated with the manufacturing process?

- Are significant amounts of water used?
- What is the water source?
- Are there treatment and/or re-use processes in use?

What kinds of mediation of ecosystem impacts are being employed?

- Are nonrenewable harvest sites being re-naturalized?
- Are farms/forests being managed appropriately?
- Are water, air, and soil mediation part of an overall management plan?
- Is protection of nearby, unaffected lands part of the management plan?

Making Good Choices

Finding the answers to all of these questions can be a daunting task, made even more difficult by the global nature of so much manufacturing. Individual components of a single construction material may involve a complicated supply chain with multiple companies and practices that may be difficult to assess. However, it is critical for those of us striving to minimize our ecosystem impacts when designing a building to ensure that we do our due diligence in this area. Too often, we only look at environmental impacts on our particular building site, and ignore the vast and interconnected webs of impacts that are woven all around the world during

Plastic foam and mycelium insulation share very similar performance traits, but the plastic foam has very high impacts for raw material extraction, high levels of air and water pollution, and creates highly toxic by-products. The mycelium product has almost no harvesting or manufacturing impacts. To achieve low ecosystems impacts in a project, each material choice should be vetted in this way. CREDIT: A: STYROFOAM FACTORY-SHUTTERSTOCK. B: MYCELIUM FACTORY - ECOVATIVE

Life Cycle Analysis or Assessment (LCA)

LCA is a mode of inquiry that attempts to quantify the environmental impacts associated with all the stages of a material or product, from raw material extraction through materials processing, manufacture, distribution, use, repair and maintenance, and disposal or recycling. The assessment follows prescriptions for creating data on a range of impacts, including:

- Fossil fuel depletion
- Other nonrenewable resource use
- Water use
- Global warming potential
- Stratospheric ozone depletion
- Ground-level ozone (smog) creation
- Nitrification/eutrophication of water bodies
- Acidification and acid deposition (dry and wet)
- Toxic releases to air, water, and land*

*from the Athena Sustainable Materials Institute, athenasmi.org

This data can help project teams accurately determine the ecosystem impacts of their design decisions and make this criterion possible to quantify, rather than relying on assumptions. A number of companies specialize in construction LCA, and their documentation is a valuable resource for project teams.

the harvesting and manufacturing of building materials.

In the next chapter, we will look at a number of material and product rating systems that can be useful in answering to these types of questions without requiring the building designer to do all of the research.

2) Embodied carbon and energy impacts

I have come to call this criterion the "carbon elephant in the room." Long disregarded as an important consideration in green building, the amount of embodied carbon in building materials is a formidable contributor to climate change; reducing embodied carbon is imperative if we are to get global greenhouse gas emissions under control. Embodied carbon is an area in which it is relatively easy to make a meaningful contribution on the scale of an individual building project, and collectively the building industry could likewise make an important global contribution by focusing on low-carbon and carbon-sequestering materials.

Throughout this book, the term "carbon" is used to indicate greenhouse gas emissions, but the scope of consideration includes all gasses that contribute to climate change. In the scientific and industry literature, the common measurement is CO_2e, which expresses the total climate change potential of emissions as a *carbon dioxide equivalent.*

Important questions to ask include:

What carbon emissions are associated with harvesting?

- What are the types and quantities of energy/fuel being used and the related emissions intensity?
- Are there non-fuel-related carbon emissions associated with the harvesting process? This could include intentional emissions such as methane venting, or unintentional emissions like the decomposition of forestry waste or releases from soil erosion.

What carbon emissions are associated with manufacturing?

- What are the types and quantities of energy/fuel being used and the related emissions intensity?
- What emissions are generated during manufacturing from non-fuel sources? This includes emissions resulting from chemical changes/processes and by-products. For example, in the manufacturing of cement, carbon dioxide in the limestone is driven into the atmosphere, contributing up to 5–7% of the world's CO_2 emissions.[5]

What carbon emissions are associated with transportation?

- Are transportation emissions being counted in the harvesting and manufacturing accounting?
- What emissions are associated with the mode(s) of transportation delivering the material from manufacturer to building site?

Is carbon being sequestered in materials?

- What is the carbon content of the material itself? Plant-based materials (wood, bamboo, cotton, straw, hemp, etc.) contain a significant amount of carbon (35–60% by weight) that has been removed from the atmosphere during the growth cycle.
- Does the carbon content in the material offset the carbon emissions released during harvesting/manufacturing?

Making Good Choices

There are a limited number of open-source carbon databases that can give general averages for the carbon associated with building materials, and these can be used to compare across material categories and establish some good guidelines about what materials have higher or lower carbon emissions. The *Inventory of Carbon and Energy (ICE) Version 2.0* is available for free from Circular Ecology, and is a good reference. There is a global standard (ISO 14064 standard for GHG emissions) for manufacturers to use to calculate their specific carbon emissions, and wherever possible it is best to use these figures, as emissions can vary widely depending on fuel source, carbon capture strategies, and manufacturing processes.

Choosing low-carbon and carbon-sequestering materials is a way to make a measurable reduction in impact. The chart below shows the embodied carbon of just the foundation, walls, and ceiling assemblies of several residential

Table 3.1-1: Embodied carbon of typical code-built house

Building Component	Code Minimum House: 1,000 sq. ft.			
	Material	Quantity	Embodied Carbon (CO_2e)*	Carbon Sequestration**
Foundation	Concrete, 8″ wall	5.9 m³/14,160 kg	1515 kg	0 kg
	Parging (1:1:6 mix)	0.3 m³/711 kg	124 kg	0 kg
	Rebar (½″)	170 kg	238 kg	0 kg
	XPS exterior insulation — 4″	9.8 m³/243 kg	1035 kg	0 kg
	Concrete, 4″ slab	9.7 m³/23,280 kg	2491 kg	0 kg
	XPS sub-slab insulation — 2″	4.85 m³/120.3 kg	512 kg	0 kg
			5915 kg	**0 kg**
Walls	2x6 framing @ 16″ OC	1.93 m³/855 kg	505 kg	1568 kg
	5.5″ fiberglass insulation	12.25 m³/367.5 kg	496 kg	0 kg
	½″ drywall interior	2.06 m³/1473 kg	574 kg	0 kg
	6 mil poly vapor barrier	95 m²/13.6 kg	35 kg	0 kg
	Latex paint	95 m²	104 kg	0 kg
	½″ OSB sheathing	1.23 m³/615 kg	609 kg	947 kg
	Vinyl siding	97 m²/413 kg	1280 kg	0 kg
Windows	PVC frames, double glaze	11.9 m²	1008 kg	0 kg
			4611 kg	**2515 kg**
Flooring	Vinyl flooring	91 m²/599.7 kg	1913 kg	0 kg
Ceiling	½″ drywall	1.18 m³/843.7 kg	329 kg	0 kg
	6 mil poly vapor barrier	95 m²/13.6 kg	35 kg	0 kg
	R-28 fiberglass	16.5 m³/495 kg	668 kg	0 kg
	Latex paint	93 m²	101 kg	0 kg
Roof	Trusses	2.36 m³/1062 kg	212 kg	1947 kg
	Metal truss plates	157 kg	229 kg	
	½″ OSB sheathing	1.5 m³/750 kg	338 kg	526 kg
	Asphalt shingles	125 m²	675 kg	
			4500 kg	**2473 kg**
		Totals	15,026 kg	4988 kg
		Net carbon footprint	**10,038 kg**	**108 kg/m²**
	The shaded materials have ***Red List*** chemical content		* CO_2e figures from ICE 2.0 database	** Carbon content of materials from ECN Phyllis 2 database

buildings; it makes clear how critically important material selection is in lowering greenhouse gas emissions.

Within the realm of conventional building, the same basic code-approved 1,000 square foot shell can produce either 22,154 lbs (10,049 kg) of CO_2e or 763 lbs (346 kg) — a 950% reduction based on a small number of material choices. The difference at the high-performance end of the scale can be even more drastic: a super-efficient home built using foam and fiberglass insulation can have CO_2e emissions of 29,881 lbs (13,554 kg)while a home using more natural materials to achieve the same level

Table 3.1-2: Embodied carbon of carbon conscious code-built house

Building Component	Code Minimum House: Low Carbon Choices			
	Material	Quantity	Embodied Carbon (CO_2e)*	Carbon Sequestration**
Foundation	Concrete, high slag, 8″ wall	5.9 m³/14,160 kg	1090 kg	0 kg
	Parging (hydraulic lime)	0.3 m³/711 kg	124 kg	25 kg
	Rebar (½″)	170 kg	238 kg	0 kg
	EPS exterior insulation — 4″	9.8 m³/243 kg	800 kg	0 kg
	Concrete, 4″ slab	9.7 m³/23,280 kg	1792 kg	0 kg
	EPS sub-slab insulation — 2″	4.85 m³/120.3 kg	396 kg	0 kg
			4440 kg	**25 kg**
Walls	2x6 framing @ 16″ OC	1.93 m³/855 kg	505 kg	1568 kg
	5.5″ cellulose insulation	12.25 m³/686 kg	73 kg	1006 kg
	½″ drywall interior	2.06 m³/1473 kg	574 kg	0 kg
	6 mil poly vapor barrier	95 m²/13.6 kg	35 kg	0 kg
	Latex paint	95 m²	104 kg	0 kg
	1.5″ wood fiber board sheathing	3.7 m³/980.5 kg	294 kg	1618 kg
	Wood siding	1.5 m³/675 kg	398 kg	1238 kg
Windows	PVC frames, double glaze	11.9 m²	1008 kg	0 kg
			2991 kg	**5430 kg**
Flooring	Engineered wood flooring	91 m²/764.4 kg	566 kg	830 kg
Ceiling	½″ drywall	1.18 m³/843.7 kg	329 kg	0 kg
	6 mil poly vapor barrier	95 m²/13.6 kg	35 kg	0 kg
	R-28 cellulose	18.9 m³/604.8 kg	64 kg	887 kg
	Latex paint	93 m²	101 kg	0 kg
Roof	Trusses	2.36m³/1062 kg	212 kg	1947 kg
	Metal truss plates	157 kg	229 kg	
	½″ OSB sheathing	1.5 m³/750 kg	338 kg	526 kg
	Asphalt shingles	125 m²	675 kg	
			2549 kg	**4190 kg**
		Totals	9980 kg	9645 kg
		Net carbon footprint	**335 kg**	**3.6 kg/m²**
	The shaded materials have ***Red List*** chemical content		* CO_2e figures from ICE 2.0 database	** Carbon content of materials from ECN Phyllis 2 database

of performance can actually *sequester* 31,303 lbs (14,199 kg) of CO_2!

Individual carbon emissions translate into truly meaningful levels when we look at the figures for the entire housing industry. In 2015 in the United States, 740,000 new single-family dwellings were built, at an average size of 2,392 square feet.[6] Here are the carbon implications of this volume of building:

- Conventional code-built shells — 19,607,176 tons of CO_2e emissions

Table 3.1-3: Embodied carbon of high performance, high carbon house

Building Component	High Performance House			
	Material	Quantity	Embodied Carbon (CO_2e)*	Carbon Sequestration**
Foundation	Concrete, 8″ wall	5.9 m³/14,160 kg	1515 kg	0 kg
	Parging (1:1:6 mix)	0.3 m³/711 kg	124 kg	0 kg
	Rebar (½″)	170 kg	238 kg	0 kg
	XPS exterior insulation — 6″	14.8 m³/367 kg	1563 kg	0 kg
	Concrete, 4″ slab	9.7 m³/23,280 kg	2491 kg	0 kg
	XPS sub-slab insulation — 4″	9.7 m³/240.6 kg	1025 kg	0 kg
			6956 kg	**0 kg**
Walls	2x6 framing @ 16″ OC	1.93 m³/855 kg	505 kg	1568 kg
	5.5″ fiberglass insulation	12.25 m³/367.5 kg	496 kg	0 kg
	5″ XPS foam	12.3 m³/305 kg	1299 kg	0 kg
	½″ drywall interior	2.06 m³/1473 kg	574 kg	0 kg
	6 mil poly vapor barrier	95 m²/13.6 kg	35 kg	0 kg
	Latex paint	95 m²	104 kg	0 kg
	½″ OSB sheathing	1.23 m³/615 kg	609 kg	947 kg
	Vinyl siding	97 m²/413 kg	1280 kg	0 kg
Windows	PVC frames, triple glaze	11.9 m²	1121 kg	0 kg
			6023 kg	**2515 kg**
Flooring	Vinyl flooring	91 m²/599.7 kg	1913 kg	0 kg
Ceiling	½″ drywall	1.18 m³/843.7 kg	329 kg	0 kg
	6 mil poly vapor barrier	95 m²/13.6 kg	35 kg	0 kg
	R-50 fiberglass	33 m³/990 kg	1337 kg	0 kg
	2″ spray foam insulation	4.7 m³/116.5 kg	496 kg	0 kg
	Latex paint	93 m²	101 kg	0 kg
Roof	Trusses	2.36m³/1062 kg	212 kg	1947 kg
	Metal truss plates	157 kg	229 kg	
	½″ OSB sheathing	1.5 m³/750 kg	338 kg	526 kg
	Asphalt shingles	125 m²	675 kg	
			5665 kg	**2473 kg**
		Totals	18,644 kg	4988 kg
		Net carbon footprint	**13,656 kg**	**147 kg/m²**
	The shaded materials have ***Red List*** chemical content		* CO_2e figures from ICE 2.0 database	** Carbon content of materials from ECN Phyllis 2 database

- Conventional carbon-conscious shells — 676,171 tons of CO_2e emissions
- High-performance, high-carbon shells — 26,446,765 tons of CO_2e emissions
- High-performance, natural materials — 27,705,292 tons of CO_2e are sequestered

The differences in these approaches are quite remarkable. Rather than being responsible for millions of tons of carbon emissions, buildings can easily be responsible for sequestering them away. When millions of tons of carbon can be kept out of the atmosphere through simple material substitutions on construction sites, clearly there is a strong need for any builder concerned about climate change to carefully consider their choices.

Table 3.1-4: Embodied carbon of high performance, low carbon house

Building Component	High Performance, Low Carbon House			
	Material	Quantity	Embodied Carbon (CO_2e)*	Carbon Sequestration**
Foundation	14″ compressed earth block	17.3 m³/27,697 kg	1690 kg	0 kg
	Parging (hydraulic lime)	0.3 m³/711 kg	124 kg	25 kg
	Hempcrete insulation — 12″	8.85 m³/2655 kg	377 kg	4222 kg
	Clay floor, 4″ slab	9.7 m³/10,563 kg	55 kg	0 kg
	Linseed oil finish for clay floor	91 m²/53 kg	1690 kg	0 kg
	6.5″ perlite insulation	15.3 m³/1150 kg	598 kg	0 kg
			4534 kg	**4247 kg**
Walls	2x4 box frames	1.67 m³/751.5 kg	443 kg	1378 kg
	Straw bale insulation — 14″	35.4 m³/4248 kg	43 kg	7009 kg
	1″ clay plaster	2.5 m³/2722.5 kg	14 kg	0 kg
	Lime paint	95 m²/90 kg	69 kg	0 kg
	1.5″ wood fiber board sheathing	3.7 m³/980.5 kg	294 kg	1618 kg
	Wood siding	1.5 m³/675 kg	398 kg	1238 kg
Windows	Alum. clad wood, triple glaze	11.9 m²	621 kg	0 kg
			1882 kg	**11243 kg**
Flooring	Hardwood flooring — ¾″	1.8 m²/1238.4 kg	248 kg	2270 kg
Ceiling	Softwood ceiling — ⅝″	1.5 m³/675 kg	398 kg	1238 kg
	6 mil poly vapor barrier	95 m²/13.6 kg	35 kg	0 kg
	R-60 cellulose	40.1 m³ /1283.2 kg	136 kg	1882 kg
	Lime paint	90 kg	69 kg	0 kg
Roof	Trusses	2.36 m³/1062 kg	212 kg	1947 kg
	Metal truss plates	157 kg	229 kg	
	¾″ strapping	0.51 m³/229.5 kg	46 kg	114 kg
	Metal roofing	125 m²	2464 kg	
			3837 kg	**7451 kg**
		Totals	10,253 kg	22,941 kg
		Net carbon footprint	**-12,688 kg**	**-136.6 kg/m²**
			* CO_2e figures from ICE 2.0 database	** Carbon content of materials from ECN Phyllis 2 database

Carbon Emissions and Sequestration

The embodied carbon of a building material is the total of all greenhouse gas emissions resulting from the harvesting and manufacturing processes. All building materials and products have some amount of embodied carbon, and this figure is expressed as $kgCO_2e/kg$ – kilograms of carbon equivalent per kilogram of material. Factors for $kgCO_2e$ are found in all embodied carbon databases.

What is much less likely to be found in a database is the carbon sequestration of a material. Plant-based materials "digest" atmospheric carbon during photosynthesis, and make up between 35–60% of the dry weight of plant materials, depending on the plant type and growing conditions. Left in the biosphere this carbon content is released again as the plant decomposes (or is burned), but if it is stored in a building it is removed from the atmosphere and is sequestered in the building. For building materials like wood, wood fiber, straw, hemp, bamboo, and cork, this can result in vastly more carbon being sequestered than being emitted during harvesting and manufacturing, as seen in the table below.

Clearly, significant reductions in carbon footprint can be achieved by replacing high-carbon materials like foam and mineral wool with sequestering materials like hemp and straw. These materials also reward the use of more insulation; adding more thermal performance using sequestering materials further reduces the embodied carbon footprint of the building while reducing the operational carbon footprint.

Table 3.2: Carbon emissions and sequestration for various materials

Insulation Material	Embodied Carbon (by Weight)*	Embodied Carbon for 4x8 Foot Wall @ R-28**	CO_2e Sequestered (-) or Emitted (+) for 4x8 Wall @ R-28
Hempcrete	0.142 $kgCO_2e/kg$	45.2 $kgCO_2e$	-87.1 kg per panel
Straw bales	0.063 $kgCO_2e/kg$	21.75 $kgCO_2e$	-78.3 kg per panel
Cork	0.19 $kgCO_2e/kg$	12.5 $kgCO_2e$	-45.9 kg per panel
Dense-packed cellulose	0.63 $kgCO_2e/kg$	41.3 $kgCO_2e$	-18.8 kg per panel
Denim batt	1.5 $kgCO_2e/kg$	22.45 $kgCO_2e$	+2.6 kg per panel
Fiberglass batt	1.35 $kgCO_2e/kg$	17.6 $kgCO_2e$	+17.6 kg per panel
Mineral wool batt	1.28 $kgCO_2e/kg$	21.75 $kgCO_2e$	+21.75 kg per panel
Expanded polystyrene foam	3.29 $kgCO_2e/kg$	37.25 $kgCO_2ve$	+37.25 kg per panel
Extruded polystyrene foam	3.42 $kgCO_2e/kg$	38.5 $kgCO_2e$	+38.5 kg per panel
	*Figures from *Inventory of Carbon and Energy* V.2	**Material densities from *Making Better Buildings*	

3) Energy Efficiency

For most homeowners and builders, this category is a central component of a greener home, and for good reasons. The environmental impacts of inefficient homes burning vast quantities of fossil fuels have been a key driver in global climate change, while rising fuel prices make efficiency an important aspect of an economically sustainable home. This aspect of green building has received plenty of study and practical implementation, and in recent years the advent of accurate energy-modeling software has made it relatively easy to achieve a desired energy efficiency outcome.

With all of the fanfare attached to energy efficiency, it can be easy for a homeowner to get lost in all the programs, acronyms, and competing claims. Almost all energy efficiency programs are based on achieving certain annual energy usage targets, and for a homeowner the key is to define targets that align most accurately with your goals.

Building codes in North America enforce minimum energy efficiency targets, and these targets are the "baseline" by which other programs rate their own efficiency measures (for example "25% better than code"). In the next chapter, a range of different energy efficiency rating systems are explored, each with a distinct place on the performance spectrum.

Important questions to ask include:

What is the overall efficiency target?

- What is the minimum efficiency standard required by the local building code?
- Do you have a specific target for surpassing the minimum requirements?
- Is there a rating system that aligns with this specific target that will help to ensure the goal is achieved?

What is the fuel source?

- Is fuel source predetermined, or will it be selected to meet overall targets/goals?
- What is the ecological impact of the chosen fuel source? Does this align with other project goals (i.e., GHG emissions, ecological impacts)?
- What is the local/regional primary energy source(s) for grid power (electricity) or distributed energy (natural gas)? What are the environmental consequences of this energy? What efficiency factor does the local grid/distribution utility have?

What is the equipment efficiency?

- Has the efficiency rating of specific equipment choices been factored into overall efficiency rating?
- Are there significant efficiency differences among competitors within the equipment range?

What is the service life of equipment?

- Does the equipment have a warranty and/or an expected service life?
- Has replacement cost been factored into the overall cost of the system?

What is the maintenance schedule and ease of use of the system?

- Is fuel input automated? To what degree? For example, a wood stove requires manual refueling every 2–4 hours; a pellet stove can require refueling daily, weekly, monthly, or seasonally depending on the size of the hopper; a natural gas appliance doesn't require any input from the homeowner.
- What is the recommended maintenance schedule for the equipment (burners, filters, duct cleaning, etc.)?
- What is the control system for the equipment (central thermostat, room-by-room thermostat, no thermostat) and how easy is it to understand and use effectively?

What equipment other than heating/cooling appliances will be used?

- What are the ventilation requirements (by code and/or by choice) and how will the ventilation system impact efficiency?
- How do choices regarding refrigeration, cooking, washing, lighting, bathing, and entertainment appliances affect overall energy efficiency?

Will lifestyle choices support energy efficiency targets?

- Are use patterns consistent with efficiency targets?
- Has an honest and accurate energy audit been performed?

Making Good Choices

A straightforward efficiency target (i.e., *"I want to be 50% more efficient than code minimum"*) may seem like an easy goal, but getting to that goal requires both knowledge and diligence to ensure that all of the above considerations are factored in and that decisions are supporting goals.

Energy modeling software can be an indispensable ally in this pursuit. A knowledgeable consultant can help to guide choices that have

Energy Modeling

In this sample worksheet, a building is modeled as though it were built to code minimum standards (56.7kWh/m²a), and compared to a target energy use of 15 kWh/m²a.

A number of changes to the building design are tested in the computer model, and the amount of heating demand reduction achieved by each potential change is noted (column 2). The modelers then make recommendations about whether or not the measure should be carried forward into the final building design (column 3).

Simulation Results

Sim. No.		Heating Energy Demand (kWh/m²a)	Heating Demand Reduction (kWh/m²a)	Measure Carried Forward?
I	**Starting Point (East-West Axis Parallel to Road)**	56.7	-	-
II	Passive House Target	15.0	-	-
	Target Reduction (Row I - II):	**41.7**		
Orientation				
-	**Starting Point (East-West Axis Parallel to Road)**	56.7	-	Y
1A	Orient East-West Axis Perpendicular to True North	56.6	0.1	Y
Window Area				
-	**Starting Point (East-West Axis Parallel to Road)**	56.7	-	-
2A	**Reduce Overhang** Reduce from 2'-0" to 1'-6" (distance between window head and overhang remains at 2'-0")	56.6	0.1	N
2B	**Increase Glazing Area on South Façade** Increase from 9.34 m2 to 12.74 m2 (i.e. four 5'-0" x 4'-6" windows, two 4'-0" x 2'-6", two fully glazed 5'-0" x 7'-0" patio doors); glazing to floor area ratio = 6%	54.4	2.3	Y
2C	**Increase Glazing Area on South Façade** Increase from 9.34 m2 to 16.14 m2 (i.e. five 5'-0" x 4'-6" windows, two 4'-0" x 2'-6", two fully glazed 5'-0" x 7'-0" patio doors); glazing to floor area ratio = 7.8%	53.4	3.3	N
Building Enclosure				
3	**Starting Point + 2B**	54.4	-	-
3A	**Floor Insulation** Increase effective thermal resistance of floor from R-10 to R-20	45.6	8.9	N
3B	**Floor Insulation** Increase effective thermal resistance of floor from R-10 to R-55	36.3	18.1	Y
3C	**Roof Insulation** Increase effective thermal resistance of floor from R-10 to R-90	49.9	4.6	N
3D	**Roof Insulation** Increase effective thermal resistance of floor from R-10 to R-140	46.6	7.9	Y
3E	**Low SHGC for E, W and N Glazing** Switch outer pane of IGU to 3mm with Cardinal lowE 272 on surface #2 (CofG U-value = 0.72 W/m2-°C, SHGC = 0.37)	56.1	-1.7	N
3F	**Windows Frames** Upgrade from Inline Fiberglass 325 series frames (U-value = 1.23 W/m2-°C) to Zola ZNC frames (U-value = 0.78 W/m2-°C)	53.5	0.9	N

measurable impacts, while avoiding those that do not. In most cases, the cost of the consultant is earned back through avoiding expensive decisions that may not contribute meaningfully to efficiency targets, while pointing to less expensive decisions that may have large impacts. A good auditor should be able to present you with a report that clearly shows the effects of a wide range of factors. (See sidebar.)

As the chart shows, only six of the 14 measures were chosen to meet the desired energy target. Some measures did not result in enough performance gain, others resulted in additional gains but were deemed too expensive for the impact they would have.

This type of "fine-tuning" is extremely useful at the design phase to help meet energy efficiency targets.

Ventilation

4	**Starting Point + 2B + 3B + 3D**	28.9		-
4A	**Upgrade HRVs** Upgrade HRVs to Zehnder ComfoAir200 (Heat recovery effiiency = 0.92, electric efficiency = 0.42 W/(m3h))	15.9	13.0	Y
4B	**Add Sub-soil Heat Exchanger** Efficiency = 50%, length 40m, depth = 1 to 1.5m	25.1	3.9	Y

Summary

-	**Passive House Requirement**	15.0
5	**Final Package** Starting Point (see sections 3) and 4) for details) **Compactness, C = 1.02** 2B - South Glazing Area Increase to 12.74 m2 3B - R-55 Floor 3D - R-140 Roof 4A - Zehnder ComfoAir200 HRV 4B - Sub-Soil Heat Exchanger	14.8
5A	**Measure 5 with Higher Compactness Factor** Starting Point (see sections 3) and 4) for details) Compactness, C = 1.42 (switch from 1 storey to 2 storeys) 2B - South Glazing Area Increase to 12.74 m2 3B - R-55 Floor 3D - R-140 Roof 4A - Zehnder ComfoAir200 HRV 4B - Sub-Soil Heat Exchanger	12.2
5B	**Measure 5A with Decreased Glazing Area and R-Values** Starting Point (see sections 3) and 4) for details) Compactness, C = 1.42 (switch from 1 storey to 2 storeys) - Decrease south glazing area to 11.04 m2 - Decrease floor insulating value to R-40 - Decrease roof insulating value to R-100 4A - Zehnder ComfoAir200 HRV 4B - Sub-Soil Heat Exchanger	14.8
5C	**Measure 5B with Reduced Ceiling Heights** Starting Point (see sections 3) and 4) for details) Compactness, C = 1.33 - Reduce floor-to-ceiling height from 9'-0" to 8'-0" - Decrease south glazing area to 11.04 m2 - Decrease floor insulating value to R-30 - Decrease roof insulating value to R-100 4A - Zehnder ComfoAir200 HRV 4B - Sub-Soil Heat Exchanger	15.0

Credit: ZON Engineering, zonengineering.com

Cost and the notion of "payback"

The cost implications of achieving a desired efficiency target are always an important consideration. However, the notion of "payback" tends to be applied to energy efficiency in a way that is inconsistent with other aspects of a home design. While it is common to wonder about the financial payback of additional insulation or better windows, the same question is never asked about house/room size, kitchen cabinetry and countertops, paint, or any other aspect of a home. Too often, the cost of a fancy countertop or luxury tub or shower enclosure outweighs the cost of efficiency measures that were deemed to have insufficient "payback." While financial payback is one useful metric for making efficiency decisions, factors like environmental impacts (what is the payback of greatly reducing your carbon footprint?), future fuel supplies, indoor air quality, and other considerations can be equally important; a single-minded focus on efficiency payback can result in a project that does not do justice to a full set of goals.

Measuring efficiency

The work of a good energy modeling consultant can provide valuable decision-making information for a homeowner, but it is important to make sure that energy use is monitored once the building is occupied. Actual energy use should be compared to predicted energy use, and troubleshooting undertaken if the real-world results deviate too widely from the model. Improper functioning of equipment/appliances and occupant behavior are the two most common reasons for poor real-world results. The Passive House Institute, developer of one of the world's foremost energy efficiency standards, has found that "different users have, even if they live in identically constructed houses, frequently very different consumption results: Deviations of ±50% from the average consumption value are not exceptional. This deviation is caused mainly by different thermostat settings during the heating season."[7] This makes it clear that it is not enough to just have efficiency targets during the design process, but that it is important for owners and occupants of buildings to intentionally contribute to efficiency.

4) Indoor environment quality

The modern human spends approximately 90% of his/her time indoors.[8] This puts the health qualities of the indoor environment at a premium, as most of us receive nearly all of our air, light, and water from indoor sources. And yet, this aspect of building design is rarely given much serious consideration. Building codes barely address the issue, and do little to enforce what standards they espouse (except in the cases of well-documented pollutants like asbestos and lead), so it is up to homeowners and designers to ensure that a building is a safe and healthy place for all occupants.

Despite its obvious importance, the pursuit of a healthy indoor environment is among the least considered goals — even in green building projects — and one of the most difficult to quantify to ensure that targets are being met.

Air

The US Environmental Protection Agency claims that the indoor air of the average American home is five times more polluted than the outdoor air,[9] and they rate breathing inside a building as one of the top five environmental risks to public health! This should be alarming, and provide ample reason to make this an important consideration for your building.

In addition to a wide range of naturally occurring contaminants (including shoes and outdoor clothing, cooking, bathroom use, and pervasive mold spores, dust, bacteria, and viruses), indoor air is in constant contact with building materials, whether they are surface finishes, ductwork, or assembly materials. The effects of chemicals embedded in building materials have received little to no regulatory oversight or study, and many known toxins (including carcinogens, endocrine disrupters, and neurotoxins) are often included in common, off-the-shelf building materials.

With all of this in mind, important questions to ask include:

What kind(s) of air exchange strategy will be used?

- There should be no question about whether or not an air exchange strategy is needed, but it is important to consider what the strategy will be (see Building Science Basics chapter).
- What is the quality of the outdoor air around the building?
- What type and quality of air filters will be used?
- How will filter changes be made part of the regular maintenance schedule of the house?

What kind(s) of moisture regulation strategy will be used?

- How might vapor retarders, moisture barriers, mechanical ventilation, and material choices

Red List Chemicals

Building materials can contain many different chemicals, yet only lead and asbestos are directly regulated to be excluded from our buildings. Builders may want to adopt the *precautionary principle* in the area of human health and chemicals in building materials: "where there are threats of serious or irreversible environmental damage, lack of full scientific certainty shall not be used as a reason for postponing cost-effective measures to prevent degradation."*

There are some chemicals, however, that do have full scientific certainty in regard to their effects on humans and the environment. The Living Building Challenge has created this *Red List* of chemicals that cannot be contained in any building attempting to meet the Challenge, and should be adopted by all builders:

- Alkylphenols
- Asbestos
- Bisphenol A (BPA)
- Cadmium
- Chlorinated polyethylene (CPE) and chlorosulfonated polyethylene (CSPE)
- Chlorobenzene
- Chlorofluorocarbons (CFCs) and hydrochlorofluorocarbons (HCFCs)
- Chloroprene (or neoprene)
- Chromium VI
- Formaldehyde
- Halogenated fire retardants (HFRs)
- Lead
- Mercury
- Polychlorinated biphenyls (PCBs)
- Perfluorinated compounds (PFCs)
- Phthalates
- Polyvinyl chloride (PVC), chlorinated polyvinyl chloride (CPVC), polyvinylidene chloride (PVDC)
- Short-chain chlorinated paraffins (SCCPs)
- Volatile organic compounds (VOCs) in wet-applied products
- Wood treatments containing creosote, arsenic, or pentachlorophenol

You can find more information at: living-future.org

* United Nations Conference on Environment and Development (UNCED). *Principle 15. Declaration of Rio.* Rio de Janeiro, Brazil, 1992.

be part of a coherent moisture regulation strategy?
- How is mold prevention integrated into the building design?

Will the building be free from Red List chemicals? Will the design team use the Precautionary Principle? (Described in sidebar, "Red List Chemicals," page 29.)
- Will you choose to avoid materials with chemicals that have questionable health effects?
- Are there chemicals or allergens that are known to affect you or other building occupants or that you would prefer to avoid?

Will lifestyle choices support air quality targets?
- Are use patterns (cleaning products, filter replacement, etc.) consistent with air quality targets?

In the following chapter, we will examine a variety of rating and labeling systems that can help to facilitate decisions that will impact indoor air quality.

Light

The value of natural light in buildings has only recently gained any kind of mainstream consideration, and most codes have only minimal requirements for glazing area and natural light: many rooms do not even require windows if electric lighting is provided. Given the importance of natural light to our bodies, adequate access to natural light should be part of your design considerations.

Important questions to ask include:

Does every room have at least one window?
- Is at least one window visible from all locations in the home?
- Are multiple windows visible from all locations in the home?
- Is it possible to have at least two windows in each room, preferably on different walls?
- Are windows blocked or shaded from the outside?

Are areas that are underserviced by windows naturally lit in other ways?
- Have light tubes, light wells, skylights, or other strategies been considered for areas not serviced by windows?

Are natural light patterns considered in the overall design?
- Have window locations and room placements been optimized for the timing of activities in the building?

Water

Good, clean water is essential for all of us, and we tend to assume that an uninterrupted supply of good water will be provided to our buildings without much thought or effort. Those in urban areas will be required to connect to municipal water systems, and those in rural or unserviced areas will expect that a productive well can be drilled on the property. Neither of these supplies is necessarily continuous nor healthy.

Municipal water systems are treated to meet approved health standards, but this does not mean that the water supply will meet *your* standards. The recent spate of lead issues in municipal water in the United States[10] is only one indicator that even major health issues in water supplies are not always addressed; beyond these, the range of tests used for municipal water do not address the complete spectrum of contaminants common in water sources, and the levels of contamination that trigger concern may also be questionable. It is quite reasonable to consider additional filtration/purification within your building.

Private wells are subject to very little testing and maintenance. Initial tests for viability typically require only a lack of major bacteria species, and do not test for other contaminants at all. In most jurisdictions, private residential water sources are never required to undergo periodic testing.

Important questions to consider include:

What is the main source of water for my home?

- Is municipal water hook-up required?
- Are drilled wells viable in my location?
- Is surface water (rivers, lakes) a viable option?
- Is rainwater collection and treatment viable?

Do I want to consider having more than one source of water?

- Is on-site water storage a viable option?
- Is a secondary water source a possibility?

What degree of water treatment is being considered?

- Is on-site filtration part of my water strategy?
- Whole-house or point-of-use filters?
- Are there particular standards I want my water to meet?
- Are there particular contaminants I want to protect against?

What type of testing regime might be considered?
Are particular water-efficiency targets being considered?

- Is there a local restriction on water use?
- Am I attempting to meet a particular efficiency standard (see next chapter)?

Making Good Choices

Indoor environment quality involves a range of measurable criteria (air and water cleanliness, natural light quantities) that require careful consideration, but there are many aspects that are more ephemeral and arguably equally important. Factors ranging from color and texture in the building, architectural proportions, furnishing, and the presence of houseplants can all contribute to creating a healthful indoor environment. Striving to meet high standards in the measurable elements is important, but so is the design of spaces that make you feel safe, healthy, and protected.

5) Waste

The amount of solid landfill waste that accompanies conventional construction is substantial. 6.56 million tons or 11% of annual solid landfill waste volume in the United States is from residential construction sites.[11] The US National Association of Home Builders performed a study showing that the construction of an average 2,000 square foot home generates around 8,000 pounds (3628 kg) of landfill waste.[12] A well-managed project can send as little as 400 pounds to landfill. These figures make a compelling case for paying close attention to reducing construction waste.

Important questions to ask include:

Are low-waste materials being prioritized?
What amount of waste is generated?

- Can the material be composted or recycled, or will it go to landfill?
- Are offcuts and leftovers inevitable? In what quantity?
- Is the material a single ingredient, or is it a composite? If it's a composite, can the components be easily separated?

About 4 pounds of waste per square foot of building are the US average for solid waste from construction. It's easy to do better than this.

- How much packaging is used with the material, and is it recyclable?

Where does the waste go?
- Can compostable materials actually be composted locally?
- Can recyclable materials actually be recycled locally?
- Can job site practices accommodate proper separation procedures?

What happens to it when it gets there?
- Does the local landfill have good practices (i.e., separation of streams, tapping of natural gas).
- Do recycling facilities have good practices?

What is the toxicity of disposed material?
- Does the material contain any toxic ingredients?
- Does the decomposition of the material release toxins to air, soil, or water?
- Does the decomposition of the material release greenhouse gasses?

Making Good Choices

There are two aspects to managing construction waste:

- *Job site construction waste* — Considers material selections based on the quantity and type of leftovers, offcuts, and waste associated with each material during the construction phase.
- *Deconstruction waste* — Considers materials selections based on the quantity and type of waste associated with the removal/replacement of each material during future renovations or deconstruction.

Both considerations should be given equal weight, as it doesn't matter to the planet or future generations if a material is bound up in a building for 10 years or 200; either way, it will end up back in the ecosystem, and its impacts will be the same. Choosing materials that have bad or questionable disposal consequences under the pretext that "it's safe while it's in the building" is simply deferring the issue for a little while.

Reducing job site waste requires forethought and organization. The construction plan needs to account for the types and volumes of waste that will be generated and also the timing of waste production. A waste management strategy must be formulated in advance, and then communicated to everybody on the construction site. If the process for dealing with different waste streams is clearly understood and easy to implement, it has a chance of succeeding. Otherwise, laziness and indifference tend to win out.

The measurement of waste streams is critical to ensuring that targets are actually reached. Volume or weight records should be noted for all materials that leave the site during the project to determine whether the system is working or requires improvements.

6) Resilience

Historically, humans don't have a great track record for considering resilience in our built environment. "We underattend to the future, we too quickly forget the past and we too readily follow the lead of people who are no less myopic than we are," according to Robert Meyer of The Wharton School at the University of Pennsylvania in response to Superstorm Sandy.[13] Most modern homes are completely dependent on outside services to function, and quickly become problematic — and even dangerous — when the "umbilical cord" to external services is cut for some reason.

Resilience tends to conjure up images of a bunker-like approach to long-term survival. There are many degrees of resilience that can be considered when designing a home, from the ability to maintain essential water, food, and heating capabilities for a day or two all the way to long-term independence.

Important questions to consider include:

What are the potential weather/climate circumstances that could affect my home?

- What is the 100-year flood plain? Is my home on or near this plain?
- What are the highest recorded and potential top wind speeds in the area?
- What are the historic rainfall/snowfall records?
- What immediate geographical features (trees, hills, streams, etc.) offer protection or potential threat?

What infrastructure is near my home?

- Are there hospitals, fire stations, police stations, and other emergency services nearby?
- Is there potentially dangerous infrastructure (dams, power stations, transmission lines,

Resilience Principles

The Resilient Design Institute offers a list of guiding principles for those attempting to design with resilience as a goal:

1. ***Resilience transcends scales.*** Strategies to address resilience apply at scales of individual buildings, communities, and larger regional and ecosystem scales; they also apply at different time scales — from immediate to long-term.
2. ***Resilient systems provide for basic human needs.*** These include potable water, sanitation, energy, livable conditions (temperature and humidity), lighting, safe air, occupant health, and food; these should be equitably distributed.
3. ***Diverse and redundant systems are inherently more resilient.*** More diverse communities, ecosystems, economies, and social systems are better able to respond to interruptions or change, making them inherently more resilient. While sometimes in conflict with efficiency and green building priorities, *redundant* systems for such needs as electricity, water, and transportation, improve resilience.
4. ***Simple, passive, and flexible systems are more resilient.*** Passive or manual-override systems are more resilient than complex solutions that can break down and require ongoing maintenance. Flexible solutions are able to adapt to changing conditions both in the short- and long-term.
5. ***Durability strengthens resilience.*** Strategies that increase durability enhance resilience. Durability involves not only building practices, but also building design (beautiful buildings will be maintained and last longer), infrastructure, and ecosystems.
6. ***Locally available, renewable, or reclaimed resources are more resilient.*** Reliance on abundant local resources, such as solar energy, annually replenished groundwater, and local food provides greater resilience than dependence on nonrenewable resources or resources from far away.
7. ***Resilience anticipates interruptions and a dynamic future.*** Adaptation to a changing climate with higher temperatures, more intense storms, sea level rise, flooding, drought, and wildfire is a growing necessity, while non-climate-related natural disasters, such as earthquakes and solar flares, and anthropogenic actions like terrorism and cyberterrorism, also call for resilient design. Responding to change is an opportunity for a wide range of system improvements.
8. ***Find and promote resilience in nature.*** Natural systems have evolved to achieve resilience; we can enhance resilience by relying on and applying lessons from nature. Strategies that protect the natural environment enhance resilience for all living systems
9. ***Social equity and community contribute to resilience.*** Strong, culturally diverse communities in which people know, respect, and care for each other will fare better during times of stress or disturbance. Social aspects of resilience can be as important as physical responses.
10. ***Resilience is not absolute.*** Recognize that incremental steps can be taken and that *total resilience* in the face of all situations is not possible. Implement what is feasible in the short term and work to achieve greater resilience in stages.

For more information, see resilientdesign.org

factories, chemical storage facilities, etc.) near my home?

What degree of resilience do I think is important?
- How many days of water independency are important?
- How many days of electrical independency are important?
- How many days of heating independency are important?
- How many days of food independency are important?

Are the sources for repairs and replacement of building components local/accessible?
- Can you fix the major structural and mechanical components you are choosing?
- Can you operate systems without outside power and assistance?
- Are systems able to be modified without imported parts and expertise?

Making Good Choices

Predicting needs for resilience is, by definition, difficult to do. Setting definitive goals is the best way to make decisions. Take into account the predictions made by reputable sources, especially regional emergency preparedness and disaster relief organizations whose own plans will tend to be thorough examinations of the likely scenarios. At the very least, resilience goals can be in line with official expectations and based on the severity and duration of past events.

Climate change presents a degree of uncertainty for any resilience planning. Changes in weather patterns can create conditions that are not reflected by past events, both in terms of storms and short-term weather events as well as long-term changes to environmental parameters. The future needs for heating and cooling, water, and food are examples of issues with major effects on a home that are difficult to assess. Fortunately, adhering to best building practices is a good place to start: high levels of insulation, solid foundations and roofs, quality windows, and minimum needs for energy inputs all contribute importantly to future resilience.

Often, resilience and durability are considered to be one and the same. However, there is an important distinction: durable materials and systems may last a long time, but resilient materials and systems are easily repaired, replaced, or modified. One may consider a furnace with a 50-year warranty to be durable, but if the expertise and parts required to fix it are not readily accessible, it does not contribute to resilience. Remember that situations that require resilience are the times when people and parts are least likely to be available, so resilience planning requires that local materials and expertise be able to solve issues and maintain functionality.

7) Occupant Input and Durability

There are many ways of thinking about the maintenance schedule and durability of your home, and none is "more right" than any other. More so than any of the other considerations in our criteria list, this has more to do with personal preference than any other factor. This category is all about the amount of interaction you choose to have with your home, in terms of both regular functionality and the frequency of repair/replacement.

Important questions to consider include:

Functionality

What degree of personal input are you willing to perform to maintain the home's functionality?
- Are you willing to perform tasks more than once per day?
- Are you willing to perform tasks once per day?
- Are you willing to perform tasks once per week?
- Are you willing to perform tasks seasonally? Annually?

- Do you never want to perform tasks to maintain functionality?

To what degree is functionality compromised if tasks are not performed in a timely manner?
- Is occupant comfort compromised?
- Is the well-being of the home compromised?
- Are backup systems required? Are backup systems automated?

Durability

What is the expected component lifespan, given location and climate?
- Does the component have a documented history of use in your region?

What is the required maintenance regime?
- Have manufacturer recommendations been examined?

What is the ease of replacement?
- Is the component straightforward to access?
- Does the replacement process compromise home functionality?

What level of performance degradation is expected?
- Is failure of the component gradual or catastrophic?

Will functionality or appearance dictate repair/replacement?
- Will aesthetically compromised but functional components require repair or replacement?

Will lifestyle choices support functionality and durability targets?

- Are use patterns, employment requirements, skill levels, and other factors consistent with functionality and durability targets?

Making Good Choices

Choices between functionality and durability can be conflicting. For example, using a wood stove for heating requires a great deal of daily input (monitoring indoor temperature, lighting and stoking fires) to be functional, as well as regular maintenance (ash removal, chimney cleaning). However, a woodstove is also a remarkably durable appliance unlikely to need replacement for generations, and any required repairs tend to be straightforward. Conversely, a thermostatically controlled heat pump doesn't require any daily input or regular maintenance from the homeowner, but it may need replacement within 10–20 years, and any required repairs will need to be done by a licensed technician. Either choice can make perfect sense for the right homeowner.

The critical factor when considering functionality and durability is to be realistic about your own commitment to interacting with the functionality and maintenance of your home. It is important that decisions about different components are not made in isolation. Having one or two daily or weekly "chores" to maintain functionality may seem feasible, but each seemingly reasonable choice can add to an overall regimen that is overwhelming. Similarly, committing to a small amount of regular interaction with your home's functionality can be rewarding and doesn't have to be cause for avoiding certain choices.

The movement toward a more sustainable built environment will not reach its maximum potential until it becomes more culturally acceptable to participate in the functionality of our homes. Our creation of ecologically problematic buildings mirrors the movement to a "maintenance-free" built environment. Many potentially transformative building systems remain sidelined because of a general unwillingness on behalf of homeowners to engage in simple maintenance tasks of any kind. Opting for some kind of maintenance regimen — whether once every 10–20 years or once a day — is opening the door to more sustainable options.

Vinyl Siding vs Clay Plaster

The concept of durability is familiar enough; we all want buildings that last a long time. But in sustainable building, that concept can be more nuanced.

In conventional construction, *durability* and *maintenance-free* have come to mean the same thing. Many products that are sold as maintenance-free have a working lifespan longer than a typical warranty period (often based on the average duration of home ownership, which is 13 years, according to the National Association of Home Builders) and are designed to require no maintenance over this length of time.

While it is laudable to reduce the amount of maintenance a home requires, the downfall of most maintenance-free materials is that they are typically difficult or impossible to repair, and become garbage at the end of their lifespan. Maintenance-free is another way of saying "won't need work until it goes in the garbage."

A good example is vinyl siding, which is sold as being maintenance-free and is assumed to be durable. However, should the siding be damaged, repair is uncommon and often impossible. And when UV and other environmental factors finally take their toll on vinyl siding, it is removed and taken to landfill.

A radically different approach can be found with clay plaster. While not as immediately water resistant as vinyl siding, any damage to clay plaster is easily repaired by wetting the area and applying a bit more clay plaster and/or clay paint. This exterior finish, while requiring maintenance every 5–10 years, can last for hundreds of years with no waste and virtually no material costs.

One solution is not necessarily better than the other; there are valid reasons for choosing either option. But it does put the concept of durability in a different light!

A: broken vinyl siding; B: clay plaster damage; C: clay plaster repair. CREDIT: CHRIS MAGWOOD

8) Material Costs

This criterion usually shows up at the top of most homeowners' priority lists and tends to dominate discussions about other criteria. In a code-regulated context and with ever-higher expectations of occupant comfort and convenience, no building can be described as cheap (even those that loudly make the claim). There is *less-expensive* and *more-expensive* building, and a convoluted path for a homeowner to navigate between the two.

Important questions to consider include:

Does your building size match your building budget?

- Quantity is a crucial cost factor; can your building be made smaller to reduce costs?
- Are you mistaking "wants" for "needs"?

Are you considering whole system costs, not just singular components?

- Are you getting estimates on big-ticket items only?
- How are you factoring balance-of-system costs into the budget?

Component Cost vs Whole System Cost

All too often, prospective homebuilders will attempt to create a rough budget estimate for their project by finding prices for the "big-ticket" items they've identified. It can seem to be a reasonable approach, and quite often it is relatively easy to find published costs for these items. But this ignores the fact that each major component of the building is just one part of an entire system.

Here are two examples that show how wrong an estimate based on a single component can be:

Table 3.3: Component Cost Vs Whole System Cost

Photovoltaic Generation (Solar Energy) 10 Kilowatt System*				**Straw Bale Walls** 1,200 sq. ft. Bungalow — 8-foot Walls**			
PV panels, 32 @ 310 watts	$9,240	Roof mount	$1,475	Straw bales, 250	$1,000	Framing lumber	$700
		Inverter	$3,025			Insulation for sills and box beams	$160
		Optimizers	$2,980			Plaster materials	$930
		Wire, junction boxes, etc.	$810			Mesh, barriers, and miscellaneous	$550
Totals	$9,240		$8,290		$1,000		$2,340
System total	$17,530				$3,340		
Percentage cost for balance of system	47.3%				70%		

*Material costs only. Figures are rounded, and may not represent actual figures in your market. Figures courtesy of Flanagan and Sun, flanaganandsun.com

**Material costs only. Windows, doors, flashings, electrical and other wall elements not included. Figures are rounded, and may not represent actual figures in your market.

Are quality, performance, maintenance and replacement costs among your considerations?

- Are you engaging in life cycle cost analysis?

Are fair labor practices and environmental standards part of your equation?

Are you supporting your local economy with your choices?

Are you balancing material and system costs with their associated labor costs?

- Are less expensive materials going to cost you more in labor?

Are you accounting for the carrying costs of your budget?

- Interest costs are often the largest single budget component over the lifespan of a building. Are you accounting for this in your financial planning?

Making Good Decisions

When budgeting for a sustainable building project, it is good to keep the "triple bottom line" in mind:

1. Social performance (People)
2. Environmental performance (Planet)
3. Financial performance (Profit)

Material choices made solely on the basis of bottom line cost are likely to ignore some obvious repercussions in terms of social and environmental factors — there is usually a reason that one option has a lower price tag than its competitors, and it is often the degradation of people and/or the planet. The values you bring to your building project should be consistent with your purchasing values.

There is endlessly conflicting advice available on the costs of building a home. It is important that you don't rely on any form of budget estimating other than getting specific quotes for the specific components you want to use in your specific design and location. Anything else is speculation, and no way to go about planning for what is likely to be the biggest financial investment you will make. Internet claims and manufacturer advertisements are obviously suspect sources, but off-the-cuff estimates or "ballpark" figures from contractors are not necessarily any more accurate. While you may use such sources to help guide the initial planning stages for your project, no plans should be finalized without a complete and accurate quotation from all of the relevant suppliers and trades to ensure that budget expectations are realistic. Enough time should be included in the design process to ensure that quotes can be obtained and adjustments made to the building plans to reflect real-world costs. Draft plans can be sent to competing material and system providers to establish which options will best suit the budget, and the drawings can then be further completed with those options inserted.

Once decisions have been made about materials and systems, the budget process should not be considered over. Refinements to the plans can offer opportunities to lower costs, while add-ons, upgrades, and features can inflate costs. The budgeting process takes place throughout the entire design-build cycle.

9) Labor Costs and Sources of Labor

Labor arrangements for custom homes can broadly fit into three categories:

1. Owner-builder, who personally undertakes most or all of the labor.
2. Owner-contractor, who manages the project and contracts some or all of the labor from others.
3. Owner-client, who hires a general contractor to carry out all phases of the construction on the owner's behalf.

The majority of homes in North America do not involve the owner at all in the planning or construction process. In 2013 73% of the 569,000 single-family homes built in the United States were built for sale on speculation, while 14% were built by a general contractor on behalf of a client. Only 7% were owner built or owner contracted.[14] Readers of this book are likely to fall into the latter two categories.

Important questions to consider include:

Who is doing the labor?

- Who is undertaking each of the distinct elements or phases of the building?
- Are there legal/code requirements for licensed practitioners?
- What is the skill level required for the task?
- Do you have the ability to undertake tasks?

Does the required expertise exist locally?

- Is labor being imported? From how far?
- Are there service or warranty issues associated with imported labor?

Are you balancing material and system costs with their associated labor costs?

- Are less expensive materials going to cost you more in labor?

What is the value of your labor input to the project?

- Do you lose income while working on the project? What is the balance between lost income and labor costs saved?

Are signed contract documents in place before tasks begin?

- Is the scope of work carefully defined?
- Are payment terms clear?
- Are processes for dealing with deficiencies well outlined?

How are costs being calculated?

- Time and materials or fixed quote?

Making Good Decisions

The process of building a home is a cooperative effort, and relies as much on building relationships as raising walls. You shouldn't underestimate the importance of personal relationships in your building project. Finding the right people to work with is at least as important as choosing the right materials and systems. Shared values and work ethics are at the heart of every successful building project; similarly, many projects that do not go well are due to sour relationships. You should ensure that you feel comfortable with everybody you hire, from general contractor to tradespeople and laborers. Clear communication, good decision-making processes, and collaboration will encourage the best possible results.

Remember that people — not materials — make buildings.

Make no mistake; the fees charged by a general contractor (GC) are usually well deserved. There are many reasons you might choose to be your own general contractor, but saving the money you would otherwise pay a GC is not a good reason. It takes many, many hours to coordinate a construction project, and you will either put in those hours yourself or you will pay somebody to do it. There is no right or wrong choice, but cost savings are not sole the grounds on which to make the choice.

GCs typically have the advantage of experience and an established network of tradespeople and suppliers; a good GC is familiar with codes and inspection requirements and understands the sequencing of a construction project. Everything a GC does can be done by an informed and motivated homeowner, but the learning curve will be steep and there will be at least a few delays, additional costs, and other hiccups to be expected from a first-time owner-builder.

At the center of good professional relationships are good contracts. Whether you are hiring a GC, a tradesperson, or an occasional laborer, write down your agreements and have them signed. The clarity offered by a clear contract is valuable for its ability to avoid the vast majority of misunderstandings as well as its role in settling them should they arise. Many contract templates are available, but you can just as easily write down your understanding of the agreement in your own language.

10) Code Compliance

There is an inherent tension between building codes and sustainability. Building codes have largely been generated as *reactive* documents; that is, they exist in reaction to problems that have arisen with buildings in the past and attempt to address issues to prevent reoccurrences. Builders concerned with sustainably tend to be looking forward, planning buildings to achieve goals that are not necessarily in the scope of most building codes. Put simply: sustainable builders often choose new materials, technologies, and assemblies that have little or no precedent in the codes. This can lead to problems, but it doesn't have to. You will, however, have to consider code compliance very carefully.

Important questions to consider include:

What elements of your building are considered "accepted solutions"?

- Does your local code recognize each material, system, or assembly you are proposing?

What elements of your building are considered "alternative solutions"?

- Does your code include a reference, appendix, or standard that covers the material or system you are proposing?
- Does your code have an "alternative compliance pathway" for handling new materials or systems?

Are there local or regional precedents for your alternative solution(s)?

- Are there nearby code-approved examples of the exact material, system, or assembly you are proposing?
- Are there nearby code-approved examples of similar materials, systems, or assemblies?
- Are there code-approved examples in the same country? Internationally?

Do you understand the "threshold of evidence" required for alternative solutions in your jurisdiction?

- What does your building department require of you in order to consider your alternative solution?
- Have you researched tests, standards, and/or codes from other jurisdictions that would be admissible?

Do you have advocates who can help you with your alternative solution application?

- Are there manufacturers, engineers, architects, or other knowledge-holders who are able to assist you?
- Are there costs associated with obtaining this support?

Do you understand the procedure for alternative solutions in your jurisdiction?

- Are there specific forms and/or steps involved?
- What is the timeframe for having alternative solutions reviewed?
- What is the appeal process should your application be denied?

Making Good Decisions

Despite plenty of anecdotal evidence to the contrary, there is no legal justification in any North American building code for denying a permit because of the use of an alternative material or system. At the same time, there aren't any building departments that would answer a general inquiry about whether or not they will permit any given alternative solution with a general answer of "yes." Permits are not granted or denied on the basis of a single material choice, but for meeting a complete set of requirements that demonstrate the viability and safety of the entire structure.

All codes include provisions for working productively with materials that are not directly recognized by code prescriptions, or materials being used in ways that are not directly prescribed in the code. In making a permit application, you must understand what aspects of your proposed building do and do not meet the prescriptions of the local code. You must also acknowledge that plan reviewers are concerned with specifics, not generalities; permits are granted or denied based on the exact details of the proposal and if the details are missing, inadequate, or in contravention of the code, a permit will not be granted even if the idea is generally feasible. In-depth advice for code considerations is given in Chapter 9.

11) Aesthetics

Every homeowner has a vision of how his or her building should look and feel. Never before have so many aesthetic options been available to homeowners everywhere; we are no longer limited to the materials and vernacular of our particular region but can pick and choose from a vast palette of options, including millions of images online.

There is no way to prescribe a particular aesthetic or to rank options in comparison to each other. However, it is important to understand how all your other goals may intersect with your aesthetic desires.

Important questions to consider include:

Does the siting of my building directly influence my aesthetic options?
- Does the property size or orientation impose aesthetic choices or limitations?
- Do municipal by-laws (height restrictions, boundary setbacks, location and size of parking areas, exterior finishing, and/or color requirements) impose aesthetic limitations?

Do the performance targets of my building directly influence my aesthetic options?
- Do passive solar options dictate building size, shape, orientation, or window sizes?
- Do energy systems (solar panels, chimneys, and exterior mounted mechanical equipment) impose aesthetic considerations?
- Do wall and roof thickness, window performance, and other energy efficiency measures impose aesthetic considerations?

Do the chosen materials impose aesthetic considerations?
- Are the materials able to create the desired shapes and massing?
- Are materials able to accept the desired finishes?

Are there regional vernaculars that offer aesthetic possibilities?
Will lifestyle choices support aesthetic choices?
- Does the shape, flow, and finishes of the building match your daily needs?
- Does the size, location, and budget of your home match your aesthetic choices?

Making Good Choices

A house is not a home unless it satisfies your personal aesthetics. It is important to keep a strong sense of your aesthetics in mind throughout the planning and design process, and to continuously monitor decisions you make about all the other criteria to ensure that your choices align with your desired aesthetic outcome.

Despite the intimidating array of finishes and "looks" it is possible to achieve, there is an enduring elegance to allowing form to follow function. For those who are striving to design a more sustainable home, the criteria outlined in this chapter will often suggest a particular aesthetic, or at least narrow the options. Buildings that are aesthetically satisfying are often those in which the structure and the materials are allowed to "speak" in a way that is direct and unencumbered by "faux" surfaces or treatments. It can be a valuable exercise to consider the aesthetic options directly offered by all of the chosen building materials and components before thinking about covering them with an additional surface treatment.

The numbers of factors determining the final aesthetics of your home are many, and there are excellent resources and skilled professionals to assist you. Seek plenty of advice, but in this criterion more than any other be sure that any advice you accept sits well with you.

Chapter 4

Rating Systems and Product Guides

If you have read the previous chapter about defining criteria and setting goals, you probably realize that making a sustainable home takes more than slapping some solar panels on the roof and a coat of low-VOC paint on the walls. To make a meaningful impact on the environment, your health, and your finances, there are many factors to be considered and balanced.

This is a lot of work, especially if you have to do it from scratch. Luckily, there are many smart and dedicated people who have worked hard to make this easier for you by creating a variety of rating systems to help you determine if and how you can meet your own personal criteria.

This chapter provides a useful overview of a number of these ratings systems. Most of these rating systems have accompanying documentation that would rival this book in length, so what follows is not a comprehensive review of any of them. Rather, it is an attempt to show you what system(s) might best fit your needs now that you've had a chance to consider the various criteria presented in the previous chapter.

We will look at two different kinds of rating systems: *whole building rating systems,* and *material and component rating systems.*

Whole building rating systems:

- Consider the performance of the entire assembly of the building, including the structural and mechanical systems
- Consider performance from multiple criteria
- Often requires verification — of the design and/or the in-situ performance — to ensure that ratings targets are met
- Often provides certification to prove that ratings targets have been met
- Often specify particular component rating systems to meet certain targets

Material and component rating systems:

- Address the performance of individual materials, mechanical devices, or appliances
- Considers performance from at least one, but sometimes multiple, criteria
- Provides certification to a manufacturer to prove that ratings targets have been met
- Exist independently of any whole building rating systems

When choosing any kind of rating system to follow, it is important to have a well-grounded sense of your own criteria and goals. If a particular system speaks to all of your goals, then it is a good match. If it only speaks to a few of your goals, there may be a better choice, or you may choose to rely only on those certain aspects of the rating system.

The tables below attempt to correlate a range of the most trusted rating systems with the criteria we have employed to determine the minimum level of performance required.

Brief Overviews

Living Building Challenge (LBC)

This standard is overseen by the Living Future Institute, and it attempts to "dramatically raise the bar from a paradigm of doing less harm to one in which we view our role as a steward and co-creator of a true Living Future."

The system is based on seven distinct "petals," each of which has stated imperatives that must be met in order to receive certifications:

- **Place** — Projects may only be built on greyfields or brownfields.

Table 4.1: Rating systems

Program or System	Ecosystem Impacts	Embodied Carbon/ Energy	Energy Efficiency	Indoor Environment	Waste	Resilience	Durability and Maintenance	3rd party Verification?
Living Building Challenge (USA & Canada)	**Yes.** Thorough requirements for sourcing with minimal impacts for all materials.	**Yes.** Require complete embodied carbon calculation. Requires purchase of offsets.	**Yes.** There are no specific efficiency targets, but buildings must be proven to be at least net zero energy.	**Yes.** Air: a chemical Red List must be followed, and questionable chemical content must be justified. Light, biophilic design and other IEQ issues addressed. No water quality standards.	**Yes.** Net positive waste imperative requires management plan and measurement.	**No.** Many of the requirements of the program promote resilience, but there are no formal requirements.	**No.**	**Yes.** Building must be occupied for one year and verification of performance is based on actual use data.
LEED for Homes (USA & Canada)	**Partially.** Some points available for sourcing some environmentally preferable products.	**No.**	**Yes.** Range of energy performance options for different levels of certification.	**Partially.** Points available for reduction of construction contaminants and operational ventilation, and for using some low-toxicity products.	**Yes.** Requires management plan and measurement.	**No.** Many of the requirements of the program promote resilience, but there are no formal requirements.	**Yes.** Durability plan is a prerequisite.	**Yes.** Verification is based on design specifications and pre-operational site visit.
National Green Building Standard (USA)	**Partially.** Some points available for sourcing some environmentally preferable products.	**Partially.** Life cycle analysis including embodied carbon can be used for whole building, assemblies or products.	**Yes.** Range of energy performance options for different levels of certification.	**Partially.** Points available for operational ventilation and for using some low-toxicity products.	**Yes.** Requires management plan and measurement.	**No.** Many of the requirements of the program promote resilience, but there are no formal requirements.	**Yes.** Durability plan is required.	**Yes.** Verification is based on design specifications and pre-operational site visit.
Active House (International)	**Yes.** Responsible sourcing and EPDs.	**Yes.** Carbon emissions part of quantitative criteria.	**Yes.** Range of energy performance options.	**Yes.** Air, light, and water quality required to be quantified.	**Yes.** Waste management to be quantified.	**No.**	**Partially.** Life cost and maintenance to be quantified.	**Yes.**
Passive House (International)	**No.**	**No.**	**Yes.** Total heating and cooling demand of <15 kWh/m^2/yr (4.7 kBtu/ft^2/yr).	**No.** Ventilation systems required, but no air or water quality standards.	**No.**	**No.** Extreme efficiency can help with resilience during power outages.	**No.**	**Yes.** Verification based on design specifications and on site measurement of air leakage.

☛

Chapter 4

Rating Systems and Product Guides

IF YOU HAVE READ THE PREVIOUS CHAPTER about defining criteria and setting goals, you probably realize that making a sustainable home takes more than slapping some solar panels on the roof and a coat of low-VOC paint on the walls. To make a meaningful impact on the environment, your health, and your finances, there are many factors to be considered and balanced.

This is a lot of work, especially if you have to do it from scratch. Luckily, there are many smart and dedicated people who have worked hard to make this easier for you by creating a variety of rating systems to help you determine if and how you can meet your own personal criteria.

This chapter provides a useful overview of a number of these ratings systems. Most of these rating systems have accompanying documentation that would rival this book in length, so what follows is not a comprehensive review of any of them. Rather, it is an attempt to show you what system(s) might best fit your needs now that you've had a chance to consider the various criteria presented in the previous chapter.

We will look at two different kinds of rating systems: *whole building rating systems,* and *material and component rating systems.*

Whole building rating systems:

- Consider the performance of the entire assembly of the building, including the structural and mechanical systems
- Consider performance from multiple criteria
- Often requires verification — of the design and/or the in-situ performance — to ensure that ratings targets are met
- Often provides certification to prove that ratings targets have been met
- Often specify particular component rating systems to meet certain targets

Material and component rating systems:

- Address the performance of individual materials, mechanical devices, or appliances
- Considers performance from at least one, but sometimes multiple, criteria
- Provides certification to a manufacturer to prove that ratings targets have been met
- Exist independently of any whole building rating systems

When choosing any kind of rating system to follow, it is important to have a well-grounded sense of your own criteria and goals. If a particular system speaks to all of your goals, then it is a good match. If it only speaks to a few of your goals, there may be a better choice, or you may choose to rely only on those certain aspects of the rating system.

The tables below attempt to correlate a range of the most trusted rating systems with the criteria we have employed to determine the minimum level of performance required.

Brief Overviews

Living Building Challenge (LBC)

This standard is overseen by the Living Future Institute, and it attempts to "dramatically raise the bar from a paradigm of doing less harm to one in which we view our role as a steward and co-creator of a true Living Future."

The system is based on seven distinct "petals," each of which has stated imperatives that must be met in order to receive certifications:

- **Place** — Projects may only be built on greyfields or brownfields.

Table 4.1: Rating systems

Program or System	Ecosystem Impacts	Embodied Carbon/ Energy	Energy Efficiency	Indoor Environment	Waste	Resilience	Durability and Maintenance	3rd party Verification?
Living Building Challenge (USA & Canada)	**Yes.** Thorough requirements for sourcing with minimal impacts for all materials.	**Yes.** Require complete embodied carbon calculation. Requires purchase of offsets.	**Yes.** There are no specific efficiency targets, but buildings must be proven to be at least net zero energy.	**Yes.** Air: a chemical Red List must be followed, and questionable chemical content must be justified. Light, biophilic design and other IEQ issues addressed. No water quality standards.	**Yes.** Net positive waste imperative requires management plan and measurement.	**No.** Many of the require-ments of the program promote resilience, but there are no formal requirements.	**No.**	**Yes.** Building must be occupied for one year and verification of performance is based on actual use data.
LEED for Homes (USA & Canada)	**Partially.** Some points available for sourcing some environ-mentally preferable products.	**No.**	**Yes.** Range of energy performance options for different levels of certification.	**Partially.** Points available for reduction of construction contaminants and operational ventilation, and for using some low-toxicity products.	**Yes.** Requires management plan and measurement.	**No.** Many of the require-ments of the program promote resilience, but there are no formal requirements.	**Yes.** Durability plan is a prerequisite.	**Yes.** Verification is based on design specifications and pre-operational site visit.
National Green Building Standard (USA)	**Partially.** Some points available for sourcing some environ-mentally preferable products.	**Partially.** Life cycle analysis including embodied carbon can be used for whole building, assemblies or products.	**Yes.** Range of energy performance options for different levels of certification.	**Partially.** Points available for operational ventilation and for using some low-toxicity products.	**Yes.** Requires management plan and measurement.	**No.** Many of the require-ments of the program promote resilience, but there are no formal requirements.	**Yes.** Durability plan is required.	**Yes.** Verification is based on design specifications and pre-operational site visit.
Active House (International)	**Yes.** Respon-sible sourcing and EPDs.	**Yes.** Carbon emissions part of quantitative criteria.	**Yes.** Range of energy performance options.	**Yes.** Air, light, and water quality required to be quantified.	**Yes.** Waste management to be quantified.	**No.**	**Partially.** Life cost and maintenance to be quantified.	**Yes.**
Passive House (International)	**No.**	**No.**	**Yes.** Total heating and cooling demand of <15 kWh/m^2/yr (4.7 kBtu/ft^2/yr).	**No.** Ventilation systems required, but no air or water quality standards.	**No.**	**No.** Extreme efficiency can help with resilience during power outages.	**No.**	**Yes.** Verifi-cation based on design specifications and on site measurement of air leakage.

☛

Table 4.1: Rating systems (continued)

Program or System	Ecosystem Impacts	Embodied Carbon/ Energy	Energy Efficiency	Indoor Environment	Waste	Resilience	Durability and Maintenance	3rd party Verification?
R-2000 (Canada)	**Yes.** Some points available for sourcing some environmentally preferable products.	**No.**	**Yes.** Requires efficiency 50% above average.	**Partially.** Requires operational ventilation and encourages use of some low-toxicity products.	No.	**No.**	No.	**Yes.** Verification is based on design specifications and pre-operational site visit.
Energy Star (USA & Canada)	No.	**No.**	**Yes.** Requires efficiency 25% above average.	**Partially.** Requires operational ventilation and encourages use of some low-toxicity products.	No.	**No.**	No.	**Yes.** Verification is based on design specifications and pre-operational site visit.
WELL Building (USA & Canada)	No.	**No.**	No.	**Yes.** Air, water, light, and all aspects of human well-being considered.	No.	**No.**	No.	**Yes.** Verification is based on design specifications and required in situ testing.
Standard of Building Biology Testing (International)	No.	**No.**	No.	**Yes.** Air, water, light, and all aspects of human well-being considered.	No.	**No.**	No.	**Yes.** Verification by qualified inspectors.

- **Water** — 100% of water supply and treatment needs must be met on site.
- **Energy** — 105% of the project's energy needs must be supplied by on-site renewable energy on a net annual basis, without the use of on-site combustion.
- **Health and Happiness** — Projects must produce a Healthy Interior Environment Plan that explains how the project will achieve an exemplary indoor environment.
- **Materials** — Projects may not use materials that contain Red List chemicals; they must account for total embodied carbon; they must use and advocate for the creation and adoption of third-party certified standards for sustainable resource extraction and fair labor practices; they must use regionally sourced materials; and they need to reduce site waste by over 90%.
- **Equity** — Projects must contribute to communities that allow equitable access and treatment to all people regardless of physical abilities, age, or socioeconomic status; and the project must provide proof of contribution to a just society.

- **Beauty** — Projects must meaningfully integrate public art and contain design features intended solely for human delight and the celebration of culture, spirit, and place appropriate to the project's function, and educational materials about the project must be provided.

The LBC is by far the most comprehensive building rating system, and differs from others not only via its higher standards, but also by requiring projects to embody the full intent of each "petal" category in order to receive certification. There are no "partial marks" or lower thresholds, though projects can choose to certify in chosen "petal" categories rather than all of them.

The LBC is unique in requiring one full year of operation of the building before certification can be given, rather than relying on computer modeling.

This is an aspirational standard that has not been met by many projects, but for builders with an interest in making the best possible buildings, it would be the one to use.

The guideline materials for the LBC are available for free. Project teams must register and pay a fee to qualify for certification.

LEED for Homes

This is the residential rating program overseen by the United States and Canada Green Building Councils (USGBC and CaGBC), and is one of several LEED systems for different types of construction.

This is a points-based system, with four different levels of certification possible: Certified, Silver, Gold, and Platinum; levels are dependent on the number of points achieved. There are nine categories in which a project is considered, and a minimum and maximum number of points to be scored in each category:

- **Integrative Process** — 2 points. Encourages integrated design teams with all participants meeting at least once during the design.
- **Location and Transportation** — 15 points. Ensures projects meet standards for land protection, density, proximity to transportation and community resources.
- **Sustainable Sites** — 7 points. Encourages appropriate rainwater management, non-invasive vegetation, and heat island reduction.
- **Water Efficiency** — 12 points. Ensures minimum water-use standards, encourages water efficiency.
- **Energy and Atmosphere** — 38 points. Rewards energy efficiency that exceeds Energy Star requirements.
- **Materials and Resources** — 10 points. Rewards responsible resource selection and waste management.
- **Indoor Environmental Quality** — 16 points. Rewards good mechanical ventilation practices, with a small number of points for low-emitting products.
- **Innovation** — 6 points. This category allows pilot projects and/or those with exemplary performance to gain extra point credits.
- **Regional Priority** — 4 points. This category allows a project to earn points that directly respond to regional priorities (water scarcity, for example).

LEED for Homes is a widely referenced standard, and some municipal jurisdictions have adopted many LEED provisions, or offer incentives for projects that meet LEED criteria.

As can be seen from the point distribution, energy efficiency accounts for 35% of the possible points and is at the heart of this program. While LEED encourages many excellent practices, it also allows projects that make just a few good choices in a criteria area to score some or even the maximum number of points, and there is no incentive within the system to go further. An example of this is in the Indoor Environmental Quality section, where low-emitting materials are worth 3 of the 16 possible points, which can be achieved while still including a lot of products in the building that are high-emitting materials.

LEED has had an important impact on the construction industry, and the language of LEED is familiar to architects, consultants, and builders alike; the existence of the program has done much to bring attention to the need for greener buildings. A LEED Platinum distinction does represent a fairly high overall environmental performance, one which only 80 homes in Canada have achieved to date.[15]

The guideline materials for LEED for Homes are available for free. Project teams must register, work with a certified rater, and pay a fee to qualify for certification.

National Green Building Standard (NGBS)

This standard arises from collaboration between the National Association of Home Builders (NAHB) and the International Code Council (ICC) in the United States to establish a nationally recognized standard definition of green building for homes. It is a points-based system with four levels of recognition: Bronze, Silver, Gold, and Emerald.

Points are awarded in seven categories:

- Lot Design, Preparation, and Development
- Resource Efficiency
- Energy Efficiency
- Water Efficiency
- Indoor Environment Quality
- Operation, Maintenance, and Building Owner Education
- Additional Points from Any Category for Exemplary Performance

The project team uses a spreadsheet to record project information, and the relevant number of points are calculated within the spreadsheet. The spreadsheet is relatively simple to fill out, but it is lengthy (with 37 different pages and hundreds of entry points). The guidebook shows how many points each entry in the spreadsheet is worth, but there are no indicators for total available points. Having been built by a code agency, the language is dense and refers to numerous outside standards and codes. As it based on the International Residential Code, those practitioners familiar with the code's numbering system will find it much easier to use than newcomers.

The system has had a relatively high uptake; over 90,000 single-family dwellings have been certified since the program inception in 2009. Only a small fraction of those homes have been certified at the gold or emerald level, however.

Though projects outside the United States can use the spreadsheet and guidebook, certification is only available within the United States.

The guideline materials and spreadsheet for the NGBS are available for free. Project teams must register, work with a certified rater, and pay a fee to qualify for certification.

Active House

The Active House standard comes from Europe, and is intended to steer the builder toward a comprehensive and affordable pathway to high-performance, low-impact buildings. Instead of points, this system is based on a minimum threshold of performance in each criterion that can then be assigned to one of four levels.

Criteria in the system fall into three categories, with three components to each:

- **Comfort**
  - Daylight
  - Thermal Comfort
  - Air Quality
- **Energy**
  - Energy Demand
  - Renewable Energy
  - Primary Energy

- **Environment**
  - Environmental Load
  - Water Consumption
  - Sustainable Construction

Each parameter is evaluated individually and mapped using the *Active House Radar diagram* that places the quantitative parameters of the system onto a plot according to the performance levels outlined in the standard as a means to help owners and the design team visualize the degree of performance they will achieve.

This is a very flexible system that allows a team to focus on criteria that are important to them, while ensuring at least a minimum standard of performance in all criteria areas. The ability for a team to be recognized for a high level of achievement in certain areas is different than the pass/fail or high/low rankings of other systems. The weighting of each criterion is equal, so participation in the program does not push the team toward an emphasis on a particular criterion.

Though the system is largely based in Europe, there have been some North American projects. An Active House label is awarded to a building after a verification process.

The specification and radar tools are available for free, and the calculation tool is available to all members of Active House.

Passive House

The Passive House standard was developed in Germany in 1996; it focuses solely on achieving specific energy use targets.

Certification in the program requires that:

- Design loads must use less than 15 kWh/m^2 (4,755 Btu/ft^2 or 5.017 MJ/ft^2) per year in heating or cooling energy
- Total primary energy (the building's consumption plus the losses/inefficiencies of the energy supply chain) consumption for all uses must not be more than 120 kWh/m^2 (38,040 Btu/ft^2; 40.13 MJ/ft^2) per year
- The building must have a very low air leakage rate of 0.6 air changes per hour at 50 Pascals of pressure difference (ACH50)

Passive House Planning Package software is used to model the home and predict its energy use, and an on-site blower door test confirms the air leakage rate.

The efficiency required to meet Passive House standards exceeds current code requirements by 75–90%, resulting in the highest possible degree of energy efficiency.

Besides energy efficiency, there are no other requirements for Passive House certification. The requirement for high-quality mechanical ventilation systems in order to meet the standard can help to provide a good indoor air quality, but beyond this indirect effect, there are no other environmental criteria involved.

A split occurred between the German and American branches of Passive House, resulting in two different groups offering certification in the United States and Canada: the *North American Passive House Network* (using the original German standards), and the *Passive House Institute US* (PHIUS, using a modified version of the original standard). PHIUS claims "more than 1 million square feet across 1,200 units nationwide"; interest and use of both certifications has been growing quickly in North America since first being introduced in 2003.

Only qualified consultants can purchase the Passive House Planning Package software, and project teams must work with a consultant to achieve certification.

R-2000

The R-2000 program is a Canadian government initiative from the early 1980s that was designed to promote energy efficiency in homes. It has developed over time to include increasing energy efficiency standards and some degree of additional environmental responsibility.

Certification in the program has the following requirements:

- Building envelope requirements for insulation, airtightness, and window performance must exceed provincial code minimums by approximately 50%.
- Space-heating, cooling, ventilation, and water heating systems must meet minimum standards.
- Water conservation targets must be met.
- A minimum number of indoor air quality "features" must be included.
- At least two "environmental features" must be included.

While R-2000 was at the forefront of sustainable building in the 1980s and early 1990s, the updates to the program have not kept up with the kinds of standards embraced by the other programs included here. A home built to this standard will exceed the environmental performance required by code minimums, but is not as thorough in other criteria areas.

The program is only open to Canadian homebuilders and homeowners. In order to qualify for certification, the homebuilder must be trained and certified in the R-2000 program. There is no fee for the project team, though the qualified builder may charge a premium for meeting the standard.

Energy Star

The Energy Star New Homes program provides a degree of energy efficiency and construction quality intended to exceed code minimums.

Certification in the program requires:

- High-efficiency heating and cooling, typically exceeding code requirements by 20–25%
- Durability measures
- Use of Energy Star appliances and lighting
- Field verification tests by third party

In order to qualify for certification, the homebuilder must be trained and certified by Energy Star, and a third-party rater employed to ensure program standards have been met.

The ventilation requirements of the program can help improve indoor air quality, but there are no other environmental criteria or considerations within Energy Star.

Energy Star certification is available in both Canada and the United States. There is no fee for the project team, though the qualified builder may charge a premium for meeting the standard.

The WELL Building Standard

The WELL Building Standard focuses solely on human health within buildings, and was formed to address the lack of focus on occupant health issues in other green building programs. The program identifies performance metrics, design strategies, and policies to be used to improve wellness of occupants.

Certification in the program requires points in the following areas of consideration:

- Air
- Water
- Nourishment
- Light
- Fitness
- Comfort
- Mind

There are compulsory preconditions in each area, and additional "optimizations" for additional points. A WELL scorecard is created for each project, and a rating of Silver, Gold, or Platinum is possible. A third-party verification process is used to ensure scores are accurate. Recertification is required every three years to ensure buildings continue to perform at their intended levels.

The WELL Standard is focused on large buildings, and there is no inclusion of single-family residences. However, build teams interested in creating a high-quality indoor environment can use the standard as a guideline in all areas that are relevant to the home.

Building Biology Evaluation Guidelines

The Institute of Building Biology and Sustainability (IBN) offers evaluation and testing guidelines for a comprehensive list of indoor environment criteria.

The following areas of consideration are considered:

- **Fields, waves, and radiation** — Includes guidelines for AC electric fields, AC magnetic fields, radio-frequency radiation, static electric fields, static magnetic fields, radioactivity, geological disturbances, sound waves, and light.
- **Indoor toxins, pollutants, and indoor climate** — Includes guidelines for formaldehyde, solvents, pesticides, heavy metals, particles and fibers, and indoor climate (temperature, humidity, CO_2, air ions, air changes, and odors).
- **Fungi, bacteria, and allergens** — Includes guidelines for molds, yeasts, bacteria, dust mites, and other allergens.

The evaluation is not rated by points or pass/fail, but uses four categories to characterize each individual criteria examined:

- **No anomaly** — Reflects natural conditions.
- **Slight anomaly** — Precautionary remediation recommended.
- **Severe anomaly** — Conditions not acceptable by Building Biology standards; biological and health problems are likely if not remediated.
- **Extreme anomaly** — International guidelines are exceeded; biological and health problems are very likely if not remediated.

The standards of the IBN are very rigorous. Results cannot be accurately anticipated at the design phase of the building; on-site testing post-construction is the only option. However, homeowners and designers can use the principles of building biology as a design reference to help ensure a finished building that will perform well in tests.

There is no certification or labeling for buildings that have been tested to IBN standards, and not all parameters must be tested. A consultant certified by the IBN must perform all testing.

To certify or not?

Pursuing certification in any green building program allows the homeowner and the project team to verify that all of the intended goals of the project are, indeed, being met. The documentation and inspections can help to ensure that all members of the build team follow through on commitments for material selection and quality of installation, as well as waste handling and other aspects of the project that may be hard to assess or monitor. Certification documents can increase the sales price for the home by providing assurance to buyers that claims for environmental performance are true.

At the same time, the project team can simply use the published guidelines for any of these standards to create their own checklists, and not necessarily follow through with the full certification process. In the end, the decision will be based on a range of individual factors unique to each project.

Mix 'n Match

Each rating system differs in the degree to which it prioritizes particular criteria. It is likely that no one system will satisfy all the provisions of your particular Criteria Matrix, and in that case it is a reasonable approach to pick and choose the elements of two or more systems to more closely resemble the kind of standard you wish to create. Adding to a particular energy efficiency standard — perhaps one that is recognized by your local municipality for grant programs, rebates, or other incentives — with aspects of environmental performance from other standards can allow you to benefit from government programs without limiting your overall ambitions.

Evolving Standards

All of these standards are in constant development, with new versions being introduced on a regular basis to reflect increasing standards and to address weaknesses or deficiencies. It is wise to check the websites of any programs that capture your interest to make sure you and your team work from the most updated version.

The listings in this book do not include the many regional green building programs that exist; they number in the dozens across the United States and Canada. There are also a great many regional programs that are focused on one particular aspect of sustainable building, such as water conservation or waste reduction. Research the programs and standards available in your region, and you may find they suit your project well and possibly come with financial incentives.

Twenty years ago, almost none of these standards were in existence. In the upcoming decade, it is likely that new ones will be introduced, and perhaps some of those in common use today will disappear. As always, due diligence will be rewarded.

Product Declarations and Certifications

There are many means by which individual building products and appliances can be rated and quantified. The market is over-ripe with ecological certification programs; many are simple greenwashing tools, but some are stringent and meaningful. Many of the best programs are relatively new, and have not yet built up large inventories of products. However, these will continue to grow and should be checked regularly as they grow in size.

There are two different streams of product certifications — product declarations and product certifications — each of which would be used differently by a project team.

Product Declarations

These certifications do not necessarily attempt to qualify whether or not a product is better or worse for the environment; they focus on transparency, informing buyers of the complete ingredients of the product. Some declaration programs also include information about the source of raw materials, whether or not recycling programs exist, and whether the product meets other environmental standards.

A product declaration requires the project team to understand the information provided and to put it into the context of the project criteria. It is entirely possible for a product declaration to contain information that would disqualify the product for use in the project; its appearance in the database does not indicate a particular level of performance, only a high degree of transparency.

Environmental product declarations (EPDs)

An EPD is an independently verified and registered document that communicates transparent and comparable information about the life-cycle environmental impact of products. The relevant standard for Environmental Product Declarations is ISO (International Organization for Standardization) 14025, where they are referred to as "type III environmental declarations." A broad range of life-cycle impacts is quantified in the EPD, including:

- Company description
- Product description and application

- Technical data, standards and certifications (material properties)
- Base and ancillary materials
- Specific health and safety information
- Environment and health during use
- Reference service use and end of life
- Life cycle analysis, including global warming, ozone, acidification and eutrophication potential
- Resource use, including all primary and secondary energy inputs and use of fresh water
- Waste categories and output flows

EPDs allow a project team to assess a product by different criteria, all of which are verified to a high degree of transparency. It is important to note that an EPD does not require full disclosure of all product ingredients, but only whether or not there are "substances of concern."

The program is relatively new, and though the parameters are international, there has been more uptake of the program in Europe and Asia to date. However, there appears to be a fair bit of momentum behind standardized EPDs, and they will likely begin to fill an important role in the marketplace in North America and globally. If there were EPDs available for every product on the market, the work of sustainable design teams would be greatly simplified.

Declare

The Declare program is a project of the International Living Future Institute (also responsible for the Living Building Challenge), and the program is marketed as a "nutrition-label for products." It delivers succinct, easy-to-understand information about the product, including:

- Life expectancy
- End-of-life waste stream (compostable, recyclable, landfill)
- Full ingredient disclosure
- Red List chemical content
- Final assembly location
- Volatile organic compound measurements
- LEED compliance

The information given on a Declare label is not as thorough as an EPD's, but it does give full disclosure of all ingredients, which can help teams trying to meet high criteria for indoor environment quality.

Material Safety Data Sheets

Laws in the United States and Canada require all manufacturers to provide Material Safety Data Sheets (MSDS) for all consumer products. The information contained in an MSDS is very helpful in ascertaining whether or not a product meets some project criteria. MSDS contain disclosures relating to:

- Product and manufacturer information
- Hazards identification — all hazards related to the use of the product on a number rating scale
- Composition/information on ingredients — a disclosure of contents that represent more than 1% of the volume of the product. Manufacturers can declare ingredients to be "proprietary" in which case the exact ingredients are not disclosed
- First aid, fire fighting, and accidental release measures, if required
- Handling and storage requirements
- Exposure controls/personal protection
- Physical and chemical properties
- Stability and reactivity
- Toxicological information
- Ecological information
- Disposal and transport requirements
- Regulatory information

This resource can be very helpful in determining whether or not products contain ingredients of concern or pose health and safety concerns in use, storage, and/or transportation.

MSDS are not rated. The reader must put the information contained into context.

MSDS are always free. Retail outlets are obliged to stock MSDS for all products they sell, and many manufacturer websites offer MSDS electronically.

Product Certifications

The majority of product certifications have a narrow focus, certifying just one aspect of a product's environmental performance. Very few certifications consider the full range of environmental and health implications over the life cycle of the product.

CradletoCradle

The CradletoCradle (C2C) Products Innovation Institute offers a certification program that allows building products to receive a scorecard rated by Bronze, Silver, Gold, and Platinum levels in five categories:

- Material Health (free from "banned list" chemicals, provides no exposure from carcinogens, mutagens, or reproductive toxins, and meets VOC-emissions testing requirements)
- Material Reutilization
- Renewable Energy and Carbon Management
- Water Stewardship
- Social Fairness

The C2C label represents the highest overall standards in the industry today. Project teams can use the scorecard to find products that match project criteria, as it is possible for a product to rate Bronze in one category and Platinum in another.

There is no cost to use the C2C database.

EcoLogo and GreenGuard

Launched by the Canadian government in 1988, EcoLogo is one of the most established and longest-running environmental labeling programs. Underwriter's Laboratories (UL) acquired the program in 2010. EcoLogo is one of the few multi-attribute labels, considering a range of life-cycle impacts for each listed product.

UL also offers the GreenGuard certification, a single-attribute rating that is focused on human health concerns stemming from materials' emissions.

While LEED and other building rating systems recognize EcoLogo and GreenGuard as a means of ensuring that products meet a minimum environmental standard, these ratings do not ensure the highest levels of performance. They are based on industry and/or government standards, and products that surpass these minimum thresholds are not necessarily exemplary, and may just be "less bad." For example, a number of spray foam insulation products are listed in both registries, despite containing flame retardant chemicals that are known to be dangerous to human health and which appear on most chemical "Red Lists." The bar for these programs is low, but there are products that go well above that bar. These ratings are best used to narrow down the field of selection, and then you can do further research to ascertain if products meet your criteria.

The product databases for both programs are free to access.

Green Spec Directory at BuildingGreen

This subscription-based service is not exactly a ratings system, but it is a thorough and well-considered database of products that have met the standards of the BuildingGreen organization, which is the publisher of *Environmental Building News*.

The Green Spec Directory provides basic information for a wide range of materials by categories. Listings for products include a brief assessment of the material, links to manufacturer website, and helpful related articles about the product category.

The product reviewers at BuildingGreen do not adhere to a particular standard when deciding whether or not to approve a product. They explain on their website: "Our team of product experts evaluates each category of building products and materials, and sets a high bar for what we consider 'greenest of the green' in that category. We set our bar at a place that is attainable for leading-edge product manufacturers who are actively tackling the key issues in that category."

This type of rating can make it difficult to assess whether or not a product is meeting any quantitative targets for your project, but it is extremely helpful as a way to find your way to products that are excellent in their category.

Single-Attribute Ratings

Many ratings programs focus on a single aspect of a material or system, and can help to determine whether or not a choice will meet a particular project criterion. There are hundreds of different programs across North America and worldwide. The options given below are exemplary programs that will have meaningful impact on the criteria they cover.

Forest Stewardship Council (FSC)

FSC is an international not-for-profit organization with branches in both Canada and the United States devoted to the protection of forests. According to their website: "We are an open, membership-led organization that sets standards under which forests and companies are certified. Our membership consists of three equally weighted chambers — environmental, economic, and social — to ensure balance and the highest level of integrity."

FSC brings a high degree of transparency and equity to their certification programs, and wood products that are FSC-certified can be relied upon to be harvested and milled to the best standards in the industry.

Pharos Project

The Pharos Project is managed by the Healthy Buildings Network, which was founded in 2000 "to reduce the use of hazardous chemicals in building products as a means of improving human health and the environment."

Pharos maintains three databases:

- **Building product library** — Combines manufacturer transparency and independent research to provide in-depth health and environmental information about a wide range of building products.
- **Chemical and material library** — Identifies key health and environmental information using authoritative scientific data for specific human and environmental health hazards, restricted substance lists, and GreenScreen List Translator scores.
- **Certifications and standards library** — Provides information on certifications and standards used to measure the environmental and health impacts of building materials, including VOC content and emissions, recycled and bio-based content, and more.

These resources give a design team the information needed to meet the highest criteria for occupant and environmental health.

Energy Star

The Energy Star program is run by Natural Resources Canada and the Environmental Protection Agency in the United States. It is an energy efficiency standard that rates household appliances and HVAC devices according to their energy use.

The program establishes targets that are above the federal government minimums in both countries, and it awards the Energy Star label to any appliance that exceeds those standards by a certain percentage (the amount varies by category).

Energy Star is pervasive in the marketplace. In some product categories, over 80% of all available models meet Energy Star standards, which indicates that it often does not set a particularly high threshold of performance. The Energy Star label does not guarantee exemplary performance.

The free, online listings of products in the online database gives comparative energy use figures, which allows a design team to select products that have the highest efficiency of all those listed by Energy Star.

WaterSense

The US Environmental Protection Agency offers this program to certify household plumbing equipment that is at least 20% more water efficient than the marketplace average while maintaining a market-average level of performance.

Actual water-use rates are included in the free online listings of WaterSense-approved products, and these can be used to ensure that a design team can select those with the highest efficiency.

Carbon Footprint Labels

There are multiple programs around the world that attempt to rate both products and manufacturers with regard to their carbon footprint. Some prominent examples include:

- CarbonCare
- CarbonFree Certified
- Carbon Neutral Certification
- Carbon Reduction Label

For projects with carbon reduction as a key criterion, it is worth exploring these options. As a relatively new kind of certification, it is difficult to know which programs will emerge as market leaders. Design teams should research these with their own criteria in mind and choose accordingly.

Chapter 5

Criteria Matrix

CLEARLY, THERE IS A LOT TO CONSIDER when setting out to achieve a high degree of environmental performance in a new home or renovation. The preceding chapters have attempted to outline a mindset for exploring sustainable options; put forward a range of criteria for consideration, and suggest some tools to ensure those criteria are achieved.

With all of this in mind, the next step is to go about setting firm goals within each criterion. These goals will then become the directive guiding all the remaining steps in creating your building, from selecting team members to choosing materials and systems and negotiating all the myriad decisions involved in a project.

The following matrix offers a methodology for goal setting that has served a wide range of clients undertaking a diverse range of projects. Using the ten criteria explained in Chapter 3, it assists in plotting a range of possible values for each.

This goal matrix is intended to be the central decision-making tool for your entire design and construction phases. As such, positioning yourself on this matrix is a process that requires careful thought and consideration (all stakeholders in the final building should contribute). There will be an ongoing tension between idealism and pragmatism as you negotiate toward ever-firmer answers. It is likely to require some soul-searching, research and, in the end, a high degree of honesty to arrive at a completed matrix that will stand up to the realities of project decision-making.

My suggestion for this goal-setting phase of your project is:

1. Give the goal matrix consideration at the very beginning of the planning process. Follow your gut instincts, your values, and whatever degree of knowledge you have to position yourself on the matrix.
2. Do initial project research based on your first version of the goal matrix. Explore materials, assemblies, systems, and potential project team members to see if your initial goals bear a close resemblance to your research findings.
3. Revisit the goal matrix after some research time, and make adjustments as you feel they may be needed.
4. Use the Criteria Matrix to begin discussions with potential project team members (see Chapter 8). This will give all parties a clear understanding of the overall intent of the project.
5. Finalize a Criteria Matrix through an integrated design process, in which all team members understand the goals, have had the opportunity to consider them, and can discuss them in a shared, coordinated meeting.

After these steps, the goal matrix should be a solid, reliable document for guiding the project through design and construction phases.

Table 5.1: Criteria Matrix

Criterion	1	2	3	4
Ecosystems impacts	**Code compliance:** No particular requirements.	**Some environmentally preferred products** *Applicable standards:* LEED certified/silver, NGBA bronze/silver, R-2000, Active House minimum thresholds.	**Majority environmentally preferred products** *Applicable standards:* LEED gold/platinum, NGBA gold/emerald, Active House mid-range thresholds.	**All environmentally preferred products** *Applicable standards:* Living Building Challenge, Active House high thresholds.
Embodied carbon	**Code compliance:** No particular requirements.	**Reduced carbon:** Footprint 25–50% lower than conventional standard. *Applicable standards:* Active House minimum threshold.	**Low carbon:** Footprint 50–90% lower than conventional standard. *Applicable standards:* Active House mid-range thresholds.	**Ultra-low, net zero, or negative embodied carbon** *Applicable standards:* Living Building Challenge, Active House high thresholds.
Energy efficiency	**Code compliance:** Meets minimum local standard.	**Reduced energy use:** On-site energy use 25–50% less than code. May incorporate on-site renewable energy and/or renewable primary energy sources. *Applicable standards:* Energy Star, LEED certified/silver, NGBA bronze/silver, R-2000, Active House minimum thresholds.	**Low energy use:** On-site energy use 50–75% less than code. Likely to incorporate on-site renewable energy and/or renewable primary energy sources. *Applicable standards:* LEED gold/platinum, NGBA gold/emerald, R-2000, Active House mid-range thresholds.	**Ultra-low, net zero, or net positive energy use:** On-site energy use 75–100% less than code. Incorporates on-site renewable energy and/or renewable primary energy sources. *Applicable standards:* Living Building Challenge, Passive House.
Indoor environment quality	**Code compliance:** No particular requirements.	**Some low-emissions and nontoxic options** *Applicable standards:* LEED certified/silver, NGBA bronze/silver, R-2000, Active House minimum thresholds.	**Majority low-emissions and nontoxic options** *Applicable standards:* LEED gold/platinum, NGBA gold/emerald, Active House mid-range thresholds, WELL silver/gold, IBN principles followed.	**No emissions, fully nontoxic** *Applicable standards:* Living Building Challenge, Active House high thresholds, WELL platinum, IBN tested.
Waste	No measurement.	**Basic management plan and sorting; minimal reduction targets** *Applicable standards:* LEED certified/silver, NGBA bronze/silver, Active House minimum thresholds.	**Comprehensive management plan; 25–50% reduction** *Applicable standards:* LEED gold/platinum, NGBA gold/emerald, Active House mid-range thresholds.	**Comprehensive management plan; 50–100% reduction** *Applicable standards:* Living Building Challenge, Active House high thresholds.
Resilience*	**Code compliance:** No particular requirements.	**Minimal resilience targets:** 3–7 days of potential self-reliance; siting accounts for major flooding/storms.	**Moderate resilience targets:** 7–28 days of potential self-reliance. *Applicable standards:* Resilient Design Institute principles inform some design decisions.	**Complete resilience targets:** Continuous potential self-reliance. *Applicable standards:* Resilient Design Institute principles and strategies fully employed.

*The number range for this criterion does not represent a quantitative value, only a preference for a more or less resilient design. The answer range will have a major impact on the complete range of design considerations. ☛

Table 5.1: Criteria Matrix (continued)

<table>
<tr><th>Criterion</th><th>1</th><th>2</th><th>3</th><th>4</th></tr>
<tr><td>Occupant Input and Durability*</td><td>No regular input required:
Homeowner does not intend to be involved with systems or material maintenance.</td><td>Annual or bi-annual input may be required:
Homeowner willing to engage in some minimal regular involvement.</td><td>Seasonal input required:
Homeowner willing to engage in seasonal chores or maintenance.</td><td>Daily or weekly input required:
Homeowner willing to engage in chores or maintenance as part of regular functioning of home.</td></tr>
<tr><td colspan="5">*The number range for this criterion does not represent a quantitative value, only a preference for a more or less active involvement in the operations of the building. The answer range will have a major impact on mechanical systems and some material and finishes decisions. Frequency of occupancy, response times from HVAC equipment, and adaptability to changing life circumstances can all be important factors here.</td></tr>
<tr><td>
Building code compliance*</td><td>Code compliance:
Full use of prescriptive pathways.</td><td>Use of referenced standards:
Code compliance includes some or full use of recognized standards, prescriptive pathway.</td><td>Some alternative compliance:
One or more major elements of the building require an alternative compliance pathway.</td><td>Alternative compliance:
A large number of building elements require an alternative compliance pathway.</td></tr>
<tr><td colspan="5">*The number range for this criterion does not represent a quantitative value, only a preference for a more or less prescriptive approach to code compliance. There may be time and cost impacts for the answer range at the higher end of the spectrum.</td></tr>
<tr><td>
Material costs*</td><td>Less than conventional production costs (<$100/sq. ft.):
Project likely to use site-harvested, local, and/or recycled materials. Cost will be a factor in all material decisions.</td><td>Equal to conventional production costs ($100-150/sq. ft.):
Project will use conventional materials or will balance use of lower-cost materials with higher-cost.</td><td>Equal to conventional custom costs ($150-250/sq. ft.):
Project will incorporate custom design elements and some higher cost materials.</td><td>High end or luxury costs ($250/sq. ft. and up):
Project will consider most or all other criteria before cost.</td></tr>
<tr><td colspan="5">*The number range for this criterion represents a general quantitative value based on current North American market averages for construction costs (not including property costs, but including development fees, permits, and profit margins). These costs are based on conventional construction models, and they reflect common priorities (such as full basement foundations and standard kitchen cabinetry). It may be possible to employ strategies that lower costs compared to conventional figures in some areas to support higher costs in other areas. The number range should be used for evaluative purposes only, and full budget estimating must be undertaken to ensure the desired budget targets will be met.</td></tr>
<tr><td>
Labor costs and sources of labor*</td><td>Entirely owner built:
Project organized by owner and uses no or minimal paid labor.</td><td>Owner is general contractor; some hired labor:
Project organized and built by owner; certain trade work contracted to professionals.</td><td>Owner is general contractor; most or all labor hired:
Project organized by owner; professionals hired for build.</td><td>General contractor hired:
Project is initiated by owner; general contractor responsible for all construction work and coordination.</td></tr>
<tr><td colspan="5">*The number range for this criterion does not represent a quantitative value. There are likely to be higher costs associated with more hired labor; however, this does not account for lost income/productivity for the owner-builder, or the potential for extended construction times, errors and omissions, or other costs that are not uncommon for inexperienced project coordinators and builders. This criterion should be thoroughly considered and calculated to ensure desired budget targets and outcomes will be met.</td></tr>
</table>

Note Observant readers may notice that "Aesthetics" is not represented on this matrix. This is not to diminish its importance as a criterion in project decision-making, but rather to reflect the reality that aesthetics is not quantifiable in any way that would be informative on this matrix. It is suggested that Aesthetics be a meta-criteria, to which all decisions in all other criteria areas are held accountable.

Using Your Map

In Chapters 10 and 11 of this book, material and system options are accompanied by "Criterion Considerations" indicating performance according to the Criteria Matrix, enabling you to make quick comparisons between particular building elements and your overall goals. This can assist in identifying appropriate material and system choices.

The degree of overlap between your overall goals and a particular material or system is a useful decision-making tool. Obviously, a close match is easy to interpret, but mismatches are also important to notice and address. In some cases, mismatches will represent intentional anomalies such as costs that you know are higher or lower than overall targets, or the need for an alternative compliance pathway in a project that is otherwise following prescriptions. In other cases, mismatches can point out inconsistencies and issues that are important to address, such as materials that have indoor environment qualities or environmental impacts that do not meet your standards. In these cases, when the decision changes, the goal changes, *or* it is accepted as an anomaly for well-defined reasons.

Your completed Criteria Matrix is an effective symbol for the whole project and can be used in meetings, on the design desk, and at the building site to reinforce the goals of the project for all involved.

The color photo section of this book displays a wide array of projects and indicates the Criteria Matrix levels that were used to guide the project.

Chapter 6

Building Science Basics

OH NO, NOT SCIENCE! the term *building science* may sound intimidating. After all, you are just trying to build yourself a decent home — is it really necessary to delve into science?

The simple answer is: Yes. If you are familiar with the basic principles of building science, you will make better decisions about your building. You will ask important questions and be able to take part in essential discussions about your home. And fortunately, it is within everybody's grasp to understand enough fundamental building science to clarify the importance of applying it to your project.

What Is Building Science?

The concept of building science as a distinct mode of inquiry is relatively new, and asks a very basic question: *How do we create comfortable and efficient indoor spaces without causing negative human health or building durability issues?*

Answering this question requires a holistic approach, incorporating elements of architecture, engineering, physics, health and environmental science, and even the social sciences. Each of these fields is capable of shedding light on some aspect of building science, and design teams armed with adequate knowledge about building science can create a plan for building that will not only be comfortable, it will do no harm.

How Is This a *New* Field?

Shelter. For most of human history, if a building kept all — or even most — of the elements at bay most of the time and did not fall down or otherwise cause direct harm, it met our expectations. Human beings have thousands of years of experience in successfully making buildings that meet these basic requirements.

It is when we add the relatively recent (and continuously evolving) notion of *comfort* that we enter new territory. We only have a few centuries of making buildings with energy systems — combustion of wood or fossil fuels — that effectively temper living conditions with any degree of smoke control via functional chimneys. We have about one century of experience with buildings that can provide the myriad conveniences of electricity. We have just over half a century of attempting to offer year-round control of temperature and humidity to a fine degree, and only a couple decades of trying to do so without using obscene amounts of fossil fuel energy. So, although it may seem that the

Buildings from the past provided adequate shelter, but do not meet modern notions of comfort.

basic question of building science should be something we've already figured out, it really is a relatively recent pursuit.

Comfort *AND* Energy Efficiency

For a brief time in the 20th century, we achieved a high degree of occupant comfort in our buildings by using fossil fuels relatively indiscriminately; we filled leaky homes with conditioned air at a rate slightly faster than it exited. We were able to reliably meet thermostat set temperatures — if we used a lot of fuel to do so.

When the price of all that fuel started to rise dramatically in the 1970s, our first reaction was to add insulation to buildings to slow down the loss of heat. However, these new, more insulated houses were still relatively leaky and did not meet predicted energy savings. By the late 1980s, we were beginning to attempt to both insulate and seal buildings against gross air leakage. These buildings were, indeed, more energy efficient. However, they also experienced many forms of moisture-related failure. The efforts to solve these failures are at the heart of building science.

The rest of this chapter will attempt to make clear the reasons for those failures and what has been learned to ensure that we can, indeed, *create comfortable and efficient indoor spaces without causing negative human health or building durability issues.*

Thinking about "Control Layers"

A well-designed building enclosure does four things:

1. Keeps water out.
2. Controls air flow into and out of the building.
3. Keeps heat energy in or out, as desired.
4. Manages vapor migration.

A building that effectively controls water, air, heat, and vapor according to the demands of the climate and the needs of the occupant is a successful, comfortable, efficient, and durable building. Your basic understand of building science begins with identifying each of these four control layers in your building.

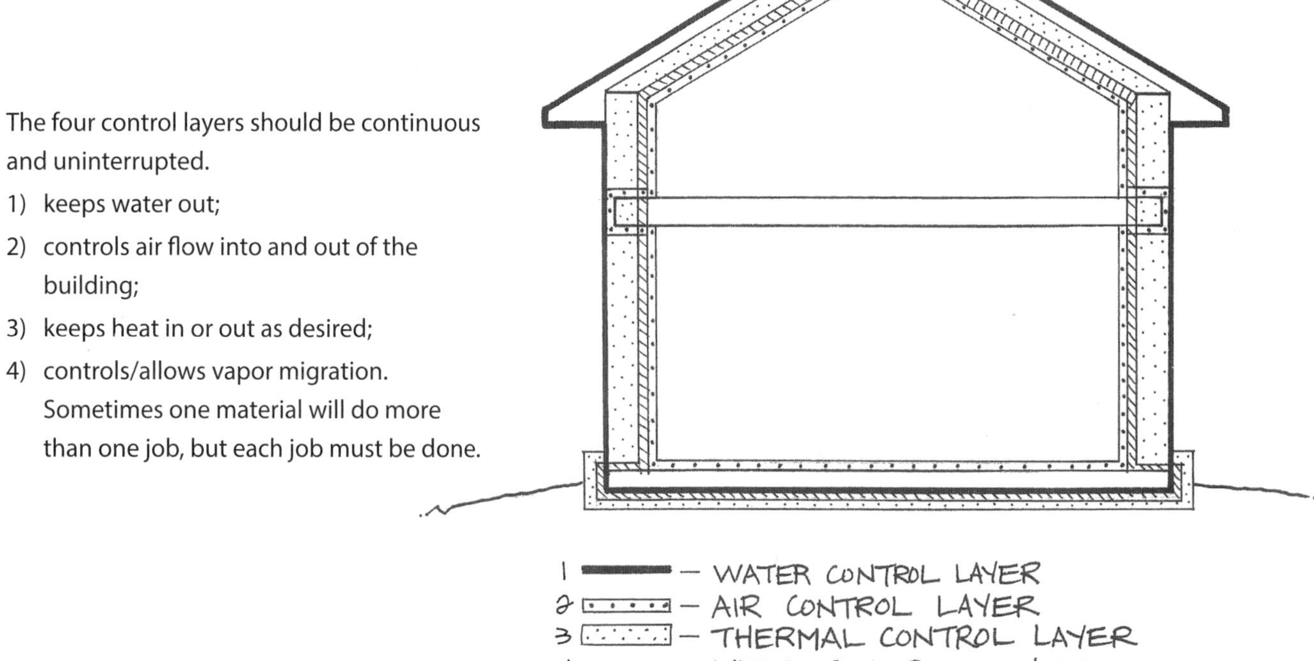

The four control layers should be continuous and uninterrupted.

1) keeps water out;
2) controls air flow into and out of the building;
3) keeps heat in or out as desired;
4) controls/allows vapor migration. Sometimes one material will do more than one job, but each job must be done.

Water Control Layer

This is the most critical of all the control layers in the building, because if vulnerable parts of the building are damaged by being repeatedly exposed to wetting, it is unlikely to meet any of our remaining criteria for a successful building.

Our buildings are exposed to water in many different forms and from every possible direction. The role of the water control layer is easy to understand. It is there to *shed and repel all water from the outside.*

However, it requires careful thought and consideration to properly design and install an effective water control layer because this layer is typically composed of a number of different materials in different parts of the building enclosure:

- Roofing and weather-resistant barrier (WRB, also called *underlayment*)
- Wall cladding and WRB
- Windows, doors, and flashings
- Foundation waterproofing
- Ground waterproofing

It is important to acknowledge that there will be many penetrations in the water control layer (chimneys and vents; window and door openings; plumbing, wiring, etc.) and that addressing water control at penetrations is a crucial part of the water control layer.

When building assemblies are exposed to excessive amounts of water, a number of serious issues can arise:

- *Mold, mildew, and rot* — Fungal growth in building assemblies can cause noxious effects on occupants and rot and decay in building materials. In the worst-case scenarios, these can lead to significant health effects and the failure of structural systems.
- *Degradation of thermal performance* — Insulation materials that are saturated with water no longer perform their function effectively, resulting in poor efficiency and reduced comfort.
- *Freeze/thaw damage* — In cold climates, water absorbed into building materials will expand when it freezes, causing cosmetic and structural damage to masonry, plaster, wood, and other porous materials. This type of damage often results in greater failure of materials to repel water, creating a cycle of increasing damage.

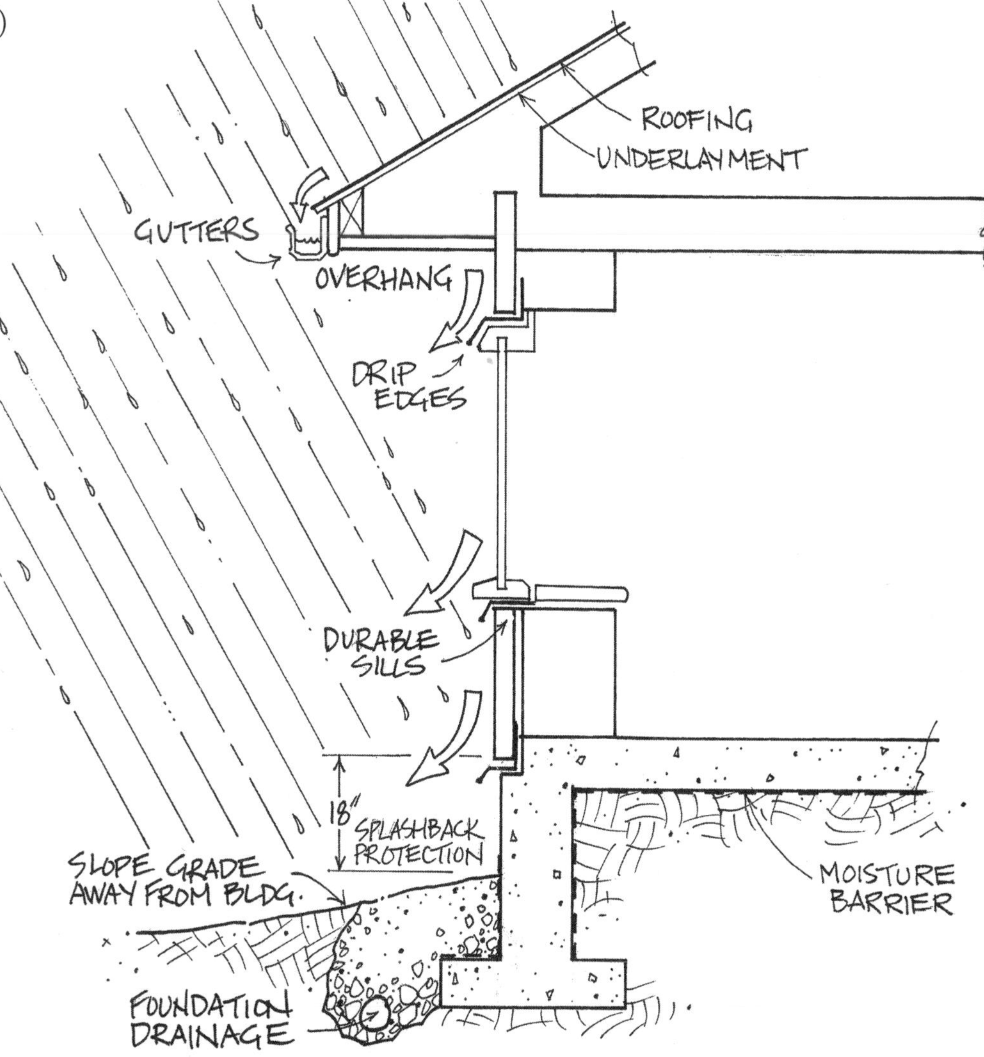

- *Expansion/contraction damage* — In any climate, water absorbed into building materials can cause swelling (and then shrinkage as drying occurs). This type of damage often results in further failure of materials to repel water, as well as cosmetic damage.
- *Oxidation and chemical decay* — The presence of water can cause oxidation decay of some metal materials. Water can also carry dissolved salts in masonry and plaster materials, causing *efflorescence* that can be merely cosmetic or structurally damaging.

Water damage in all of these forms is the leading cause of building failure, and this puts a premium on ensuring that you design and install a proper water control layer for your building.

Key Concepts for the Water Control Layer

In order to ensure that your water control layer functions as intended, there are a few important concepts to understand:

- **Waterproof versus water resistant** — A water control layer might employ materials that are:
  - Waterproof — Closed-pore materials that will not allow any water to pass through them. Waterproof materials (such as metal, rubber, most plastics, and glass) are commonly used in the areas in which complete water exclusion is crucial and/or where control is likely to be required frequently or constantly. When employing waterproof materials, it is crucial to think about the implications of water being trapped on the "wrong" side of the material and being unable to dry out.
  - Water resistant — Porous materials that can deflect and/or absorb water are often used in a water control strategy for areas of a building that receive intermittent wetting. These porous materials can be used as a primary element of a water control layer where climatic conditions and design details allow (that is, where periods of wetting are followed by sufficient periods of drying).

  Porous materials can also be treated with permeable or impermeable paints and stains to enhance their performance as a water control layer (as is often the case with wood cladding).

Most water control layers are composed of a mixture of waterproof and water-resistant materials, employed to meet the needs of a particular climate in a particular position on the building.

- **Exposed and hidden layers** — Although we tend to refer to the water control layer in the singular, its importance often warrants two (or more) layers of protection. This is common in modern construction on the roof, where the water control provided by the *roofing material* is backed up with a water control layer in the form of roof *underlayment,* and on walls, where the *exterior cladding* is often installed over a *weather resistant barrier (WRB).* These hidden layers of water control are often critical components in maintaining water protection. They are especially important for providing protection at punctures, penetrations, and transitions such as windows and skylights because they provide a surface for adhering tapes, caulking, and other sealants that protect those joints from direct weather and UV exposure.
- **Sealing of openings/penetrations** — Punctures in the water control layer (such as an electrical service entrance) must be properly flashed or sealed. Modern construction offers a range of tapes, sealants, and membrane materials that can be used to ensure water is kept out at these vulnerable locations.
- **Using caulk wisely** — Caulking is often used as an important part of the water control layer, and if a quality caulking is properly applied in the right conditions, it can play an effective

role in the water control layer. Caulked joints should be sized properly — not too small, and if large, a backing rod should be used. A high-flexibility and UV-resistant caulk is preferable. Always ask yourself what happens to water when it makes its way past the caulking, and be sure the answer is adequate enough to trust with the long-term health of your building.

- **Positive lapping** — The elements of a water-resistant layer are always installed in a "shingle lap," with materials lower in the assembly covered by materials higher in the assembly. Bulk water is encouraged to continue to run off at seams, rather than behind the next layer.
- **Drip edges** — While positive lapping encourages water to be shed, the phenomenon of *surface tension* can cause water to move against gravity behind lapped joints. Drip edges are flashings that cause water to bead and fall away from the building instead of creeping in behind. They can also be incorporated into parts of the assembly, such as window sills or trim.
- **Prevent capillary transmission** — Moisture barriers are employed between the soil and foundation materials and between foundation materials and walls and flooring, to prevent water movement into assemblies via capillary action.
- **Design for water control** — Roof overhangs, gutters, downspouts, site grading, and foundation drainage, while not directly part of the water control layer, are critical components of a water control system because they minimize the amount of water that needs to be controlled in more vulnerable areas.
- **Design flood-smart** — Design your building to account for known flood plains and predictions for 100- or even 200-year flood events. All the best water control strategies are useless if your building is literally under water.
- **What happens if/when it fails?** —Water always "wins" entry into our buildings in the long run. Our best efforts to control it are relatively short term, measured in years or decades. Protect the most vulnerable and least visible areas of your building with the best strategies, and consider the effects of a breach in your water control layer. Are the results catastrophic, or is there a way for water to drain and/or dry out before causing damage?

Putting It into Practice

At the design phase, you can choose to eliminate potential water control problems by creating plans that minimize trouble spots:

- Always think about the "Five Ds" of moisture control: *Design, Deflection, Drainage, Deposit, Drying.*
- Use simple roof lines that eliminate transitions (especially valleys).
- Incorporate generous roof overhangs.
- Consider rainscreen cladding (see sidebar).
- Place windows at the outer edge of the wall assembly and include a drip sill.
- Design walls (or at least cladding) that overhang the foundation, even slightly.
- Position penetrations in protected areas, and flash/seal them well.
- Protect wall-to-foundation and ground-level transitions with appropriate grading, drainage, and use of durable materials.

You should be able to trace your finger around the water control layer on your building plans and clearly understand how water is being directed and controlled on each surface, at each transition between surfaces, and especially at penetrations. Think like a water drop! As you are considering the effectiveness of your planned water control layer, remember that water will move by gravity, but it will also be driven by wind, and it will move via capillary action and surface tension in every direction. If at any point you are uncertain about the path of water under different conditions, redesign your water control layer until you are confident it will be effective.

At the construction phase, inspection of the water control layer is critical, and these inspections should be carried out by a person with direct responsibility for the warranty of the building. Major seams are obvious places to check, but many leaks occur at penetrations, so flashing details need to have positive lapping and appropriate sealants. Photo records of important details can prove very useful in helping to troubleshoot any problems that occur in the future. Be sure to sequence construction such that moisture- and temperature-sensitive materials and sealants can be installed and cured according to specifications.

Rainscreen Cladding

A rainscreen is the most durable and resilient means of cladding a building in a rainy climate (or even in dry areas that occasionally experience heavy, driving rain). Rainscreen cladding is installed so that there is an air space between the main wall assembly and the cladding. The space is typically ¾–1 inch (1.9–2.5 cm).

This air space prevents moisture-laden cladding from touching the wall and provides a pathway for drying air to carry moisture away from both the wall assembly and the cladding. The air space must be protected from intrusion by insects or animals at the top and bottom of the wall, and the air space must connect to the outside at the base of the wall and at the top be vented to the outside or into a well-ventilated soffit.

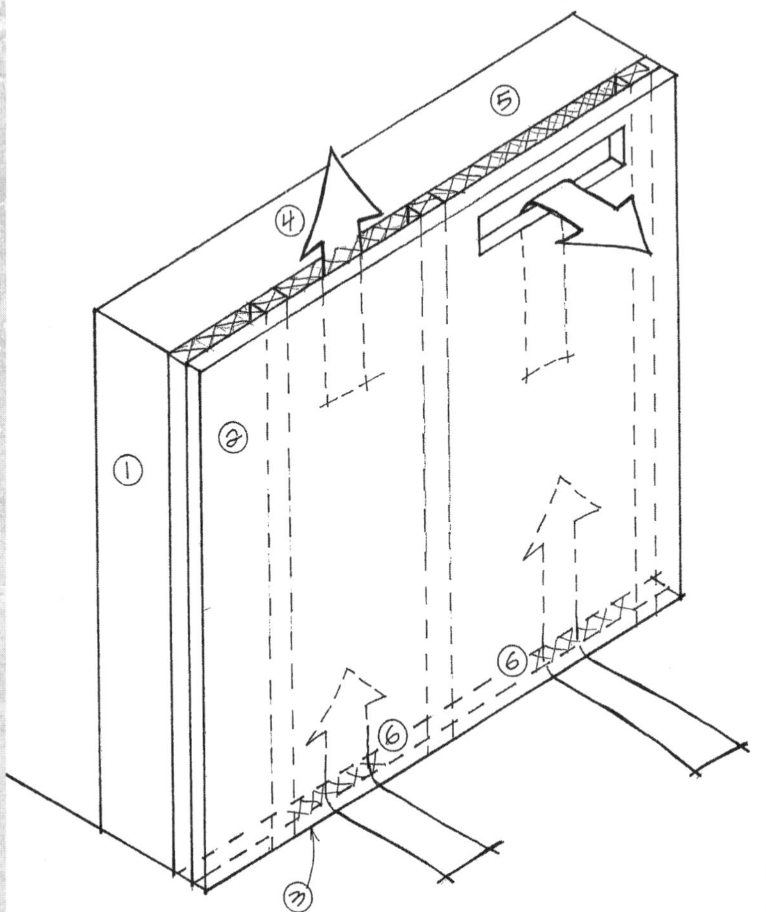

Brick, stone, wood, metal, and even plaster cladding can be installed as a rainscreen.

1) Wall assembly.
2) Siding.
3) Ventilation channel with screen.
4) Ventilation air directed into soffit or roof space.
5) Ventilation air directed outward.
6) Ventilation air enters at base of wall.

Thermal Control Layer

This layer is of primary importance for maintaining thermal comfort in the building at a desired degree of efficiency. This layer is often referred to as the *insulation* layer.

Buildings are constantly exposed to temperature differentials between an ideal indoor temperature and outdoor conditions, and the role of the thermal control layer is to *mitigate heat losses and gains through the building enclosure to maintain a consistent desired indoor temperature with minimal input from fuel sources.*

As with the water control layer, the thermal control layer is typically composed of a variety of different systems and materials that are designed to meet the particular needs of the climate and conditions in a particular location in the building.

Why Is the Thermal Control Layer Important?

When exterior conditions are far from the desired indoor temperature, a number of serious issues can arise:

- *Discomfort and occupant health* — If indoor temperatures are too low or too high, or are inconsistent within a building, occupants can experience discomfort or even ill health effects.

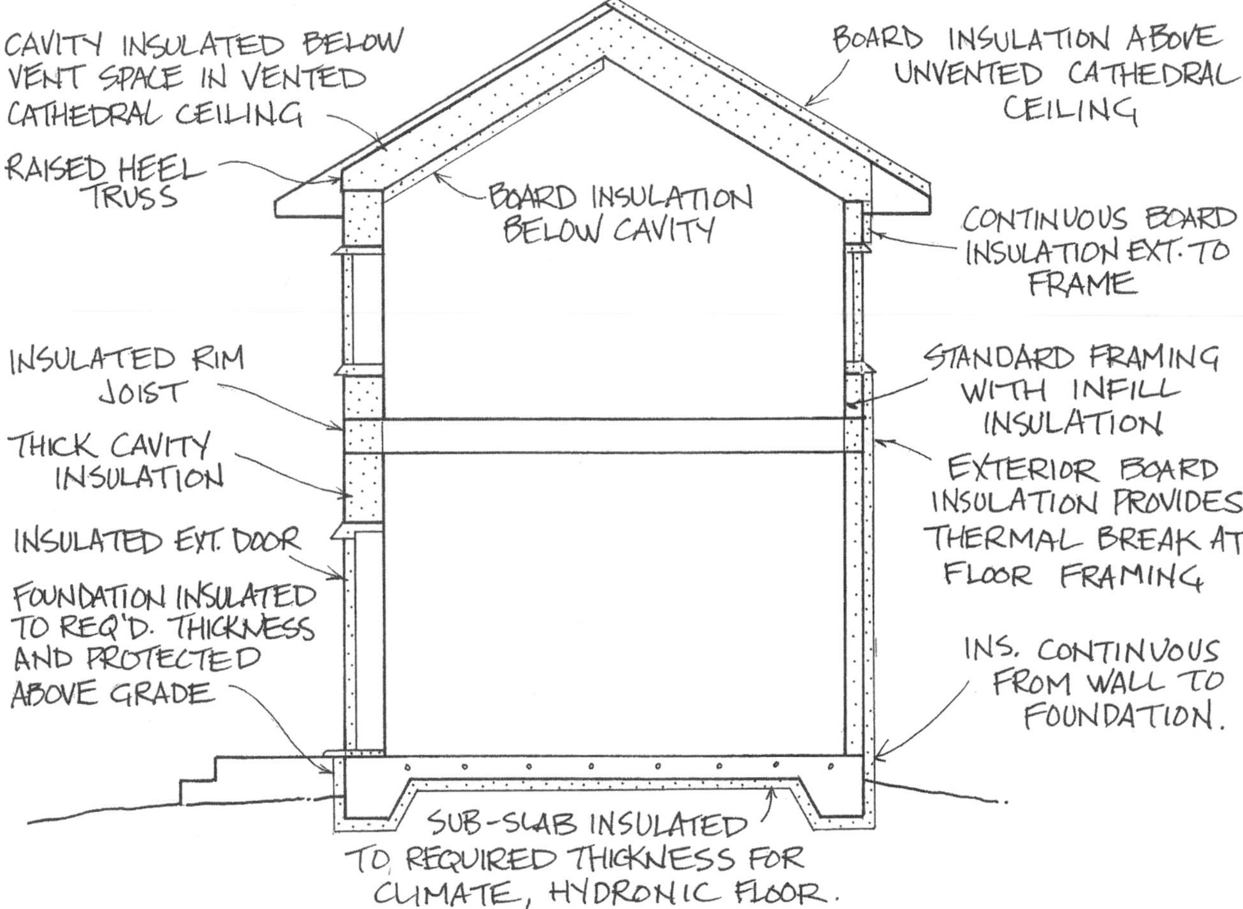

The thermal control layer is where heat energy from inside or outside the building is slowed (it is never truly stopped) by insulation, which can be located to the interior, middle, or exterior of the assembly.

- *Energy consumption* — The amount of energy required to maintain an adequate indoor temperature may be very high if the thermal control layer is inadequate. This results in fuel consumption that can be expensive and damaging to the environment.
- *Building safety* — Freezing temperatures within a building can cause plumbing pipes to burst, causing major damage to the building. Excessively high or low temperatures inside a building can compromise other building services, including refrigeration, lighting, and HVAC systems.

Key Concepts for the Thermal Control Layer

In order to ensure that your thermal control layer functions as intended, there are a few important concepts to understand:

- **There are three methods of heat transfer** — Sensible heat is that which can be measured by a thermometer, and it moves in three ways:
  - *Conduction* — The transfer of heat energy directly through a material. As heat energy from the source excites the molecules in the adjacent region of the material, they in turn excite all connected molecules. This transfer occurs in any direction, and the rate of heat flow is based on the material's conductivity, its thickness, and surface area, and the difference in temperature on either side of the material. The thermal control layer is primarily there to control conduction.
  - *Convection* — The transfer of heat through a fluid (including air). Heated fluids become less dense and rise, creating *convection loops.* The *air control layer* (see below) controls convection.
  - *Radiation* — The transfer of heat through air or space. Radiant energy moves equally in all directions, and only in straight lines.
- **Difference in temperature** — Heat moves from areas of higher concentration to areas of lower concentration; that is, from warmer places to cooler places. The rate at which heat moves is influenced by the difference in temperature, known as the *ΔT (delta T).* If it is significantly colder on the exterior of the building, there will be a stronger "drive" to the heat flow than if temperatures were similar.
- **Insulation** — Controlling the conductive flow of heat through the building enclosure requires the use of materials that are good insulators; that is, materials that *resist the conductive flow of heat.* Every surface of the building that is exposed to the exterior will require some degree of insulation to prevent conductive flows.
- **U-value and R-value** — Measuring and expressing the effectiveness of an insulation material is done in two ways. The *conductivity* of a material is expressed as a *U-factor,* and the lower the number the more resistant the material is to the conduction of heat (that is, lower numbers are better insulators). The *thermal resistance* of a material is expressed as *R-value,* and the higher the number the more resistant the material to the conduction of heat. R-value is calculated as the inverse of U-value: (R = 1/U).

  Testing of these values is typically carried out at a relatively warm temperature (70°F [21°C]) and in static conditions, with no air movement. It is important to remember that the performance of many insulation types degrades as the temperature drops and air moves around or through the material.
- **Thickness** — The effectiveness of insulation is proportional to the thickness of the material. R-value is commonly expressed as a value per inch of material thickness (R/inch).
- **Area** — The effectiveness of insulation is proportional to the surface area the insulation covers.

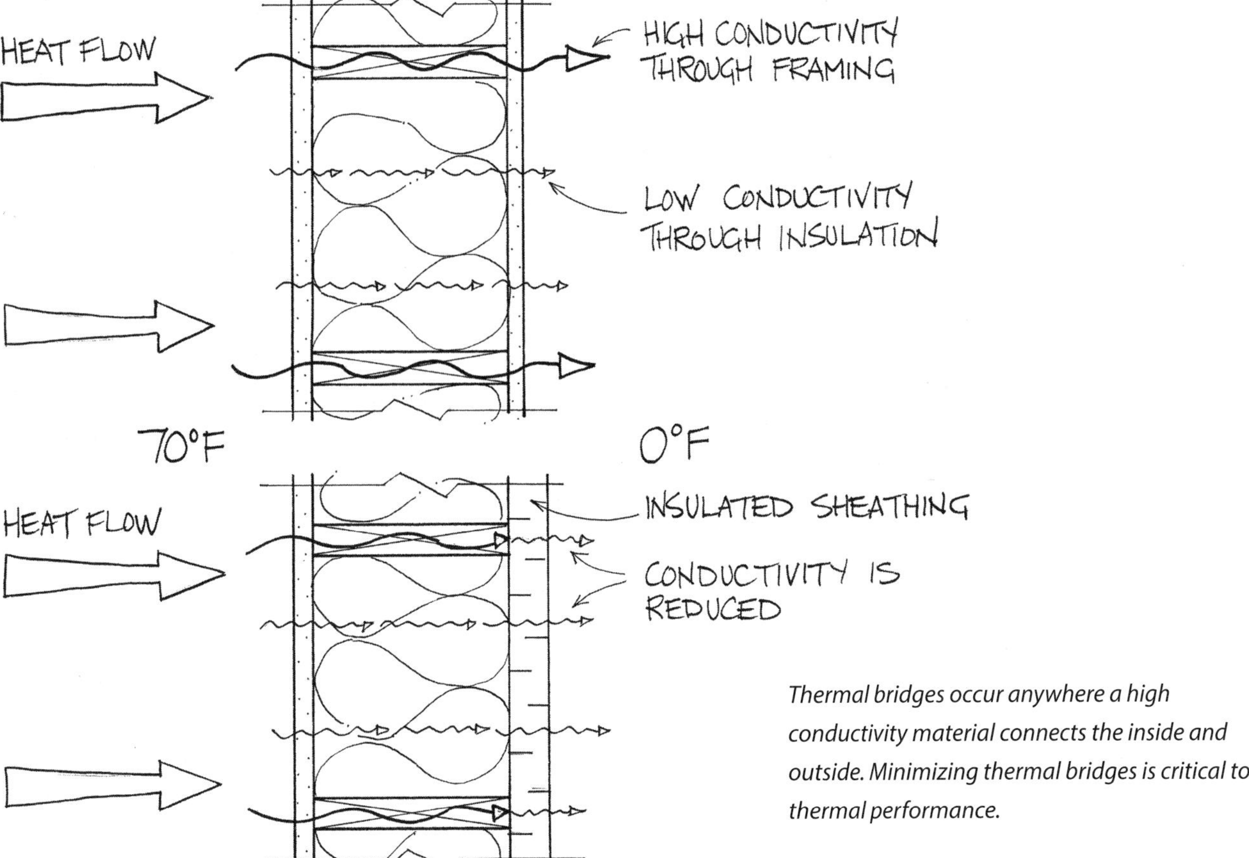

Thermal bridges occur anywhere a high conductivity material connects the inside and outside. Minimizing thermal bridges is critical to thermal performance.

- **Continuity and thermal bridging** — The overall insulation value of a building enclosure is determined by the thermal conductivity of the entire surface area of the building. If there are *thermal bridges* in the building enclosure, the increased heat movement in those areas reduces the overall thermal performance proportional to the surface area of the thermal bridges.
- **Windows and doors** — Openings (*fenestrations*) in buildings are an integral part of the thermal control layer, representing large areas of reduced insulation value. There are significant differences in thermal performance between competing windows and doors, and the greater the ΔT in the installation environment and the larger the overall area of fenestration, the greater the effect on the overall performance of the building's thermal control layer.
- **Effects of convection** — The transfer of heat by convection can have a significant impact on the performance of insulation if air is able to travel freely inside or around the insulation. To eliminate poor performance due to convection:
  - *Ensure a continuous air barrier* (see Air Control Layer, page 72) to prevent heat loss due to air movement through the thermal control layer.
  - *Install insulation materials so there are no gaps or voids* between the insulation and other elements of the enclosure. This can be particularly challenging when installing batt or rigid insulation around interruptions like framing, wiring, blocking, and other obstructions.

- **Effects of radiation** — Radiation, especially from direct exposure to sunlight, can have a significant impact on thermal performance. In areas where radiant gains from the sun need to be controlled, the thermal control layer may include:
  - *Shading* — Radiation only moves in straight lines, so blocking sunlight from striking the building is a valuable control strategy. Siting for shade from trees and buildings and shade from roof overhangs (see Passive Solar section later in this chapter) are examples of strategies you can use.
  - *Reflection* — Radiant barriers made from highly reflective materials can reflect a significant amount of heat away from the building enclosure. Light-colored roofing and reflective barriers in attics are common and effective reflection strategies.
  - *Windows and doors* — Fenestrations are particularly susceptible to unwanted radiative losses and gains. Low emissivity (low-e) coatings can help to control this.
- **Energy modeling** — Using appropriate climatic and building data can help to guide decisions regarding the design of the thermal control layer. During the modeling process, it is possible to see the effects of using more or less insulation in various parts of the thermal control layer, and to choose R-values that result in the desired outcomes for energy consumption. Energy modeling allows you to "tune" your thermal control layer for both quality and cost, to help you decide how best to meet your performance requirements within your budget.

Putting It into Practice

At the design phase, the thermal control layer requires careful thought to reduce or eliminate thermal bridges and to ensure continuity of the layer at junctions between different building components. In particular, be cognizant of:

- **Structural components and connections** — Structural elements of the building require

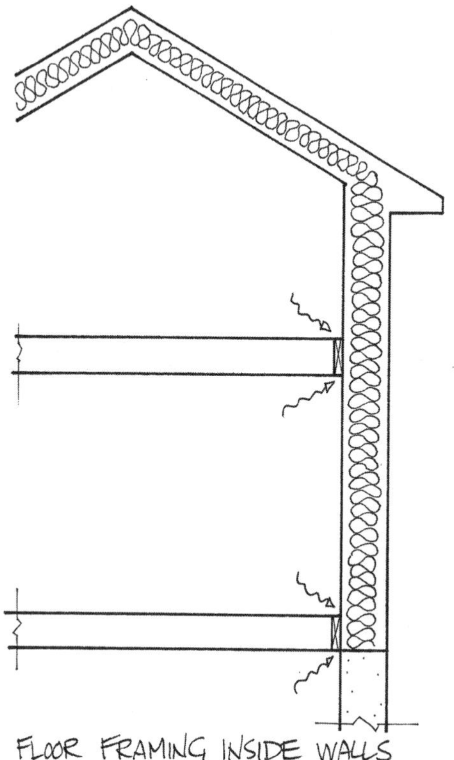

FLOOR FRAMING INSIDE WALLS
LOWER THERMAL LOSSES
+ EASIER TO AIR SEAL.

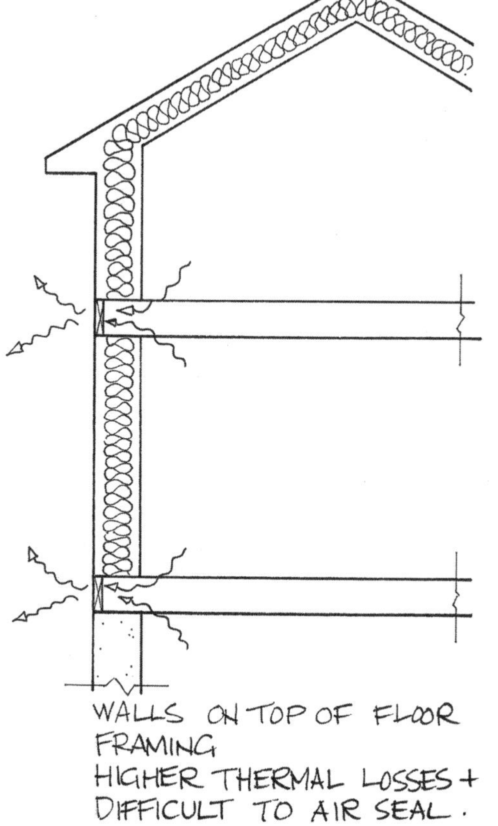

WALLS ON TOP OF FLOOR
FRAMING
HIGHER THERMAL LOSSES +
DIFFICULT TO AIR SEAL.

Floor framing systems touching the outside of the enclosure can be a point of major thermal losses and air movement.

proper detailing to avoid major thermal bridges. Wherever possible, avoid:

- *Structural elements passing from inside through to outside* — Beams, rafters, and joists are often designed to move from one side of the enclosure to the other, and these will be significant thermal bridges.
- *Floor framing systems touching the outside of the enclosure* — It is common practice to situate floor framing on top of walls, creating significant thermal bridges.
- *Bridging at window and door framing* — The frames that support windows and doors can provide a great deal of thermal bridging.
- *Bridging at wall-to-roof transitions* — A lot of structural framing combined with the angle of the roof framing can present thermal bridging issues and issues with thickness of insulation.

- **Adequate insulation from the ground** — The thermal control layer should extend to protect the building enclosure from the ground. Two lines of faulty reasoning often result in inadequate thermal control from the ground:
  - *"Heat rises"* — Hot fluids moving via convection do rise, but conductive and radiant heat flows move equally (and significantly!) in any direction. A floor or wall in contact with the ground and lacking adequate thermal control will allow very significant amounts of heat to pass into the ground. The earth is a very large thermal mass; it is able to continuously absorb heat from the building unless the thermal control layer prevents this flow.
  - *"The ground is warm"* — Surface ground temperatures vary daily and annually, based on solar exposure and air temperature, but subsoil temperatures are relatively constant and range from about 40°F (4.5°C) in the North to 75°F (24°C) in the South. While these temperatures may be closer to the comfort range than air temperatures, if the desired temperature inside the building is higher or lower than the subsoil temperature, a significant amount of energy will be required to maintain the desired indoor temperature due to losses or gains from the ground.

At the construction phase, quality of installation is critical to the performance of a thermal control layer. Even if the layer is well designed, if the installation is incomplete or of poor quality, the results can dramatically underperform. During installation, ensure:

- **No gaps or voids** — Be sure that insulation fills all cavities and makes complete, uninterrupted contact with interior and exterior enclosure elements. This can be challenging, especially when installing batt or board-style insulation around obstacles like wiring, plumbing, and blocking.
- **Proper density** — R-value of insulation materials is rated at a particular density. If the insulation is too dense (over packed), the R-value may be lower than anticipated because the air pockets in the material are compressed or eliminated. If the insulation is not dense enough (under packed or incompletely installed), the performance may be lower than anticipated because the greater volume of air spaces will allow for high losses due to convection.
- **Continuity** — A thermal control layer often combines different materials throughout the building enclosure, and it is crucial that intersections between elements of the thermal control layer are seamless, especially at crucial junctions like windows, sill plates, and wall-to-roof transitions.

Your thermal control layer is a fundamentally important element in your project. By wisely incorporating all of the above considerations to meet your specific design goals, you can closely determine the comfort and efficiency of your building.

Air Control Layer

The importance of an effective air control layer cannot be underestimated, but unfortunately it is often disregarded or misunderstood. Without an effective air control layer, your building is unlikely to meet your criteria for comfort, energy performance, or durability.

The movement of air is constant in and around our buildings. The role of the air control layer is to *prevent the uncontrolled transfer of indoor and outdoor air in the building enclosure.*

The air control layer is typically composed of a variety of different materials that are designed to meet the particular needs of the climate and conditions in a particular location in the building enclosure.

Why Is the Air Control Layer Important?

When air moves through the building enclosure, a number of serious issues can arise:

- **Occupant health** — Uncontrolled air exchange through the building enclosure results in the infiltration of air carrying particulates (dust, pollen, vehicle exhaust), chemicals (off-gassing from building materials or adjacent garages), and organic content (mold, bacteria, viruses, and excrement from insects and rodents) into the building. This contaminated air can significantly impact the indoor air quality, with mild to severe health effects.
- **Discomfort** — Uncontrolled exchange of indoor and outdoor air results in occupant

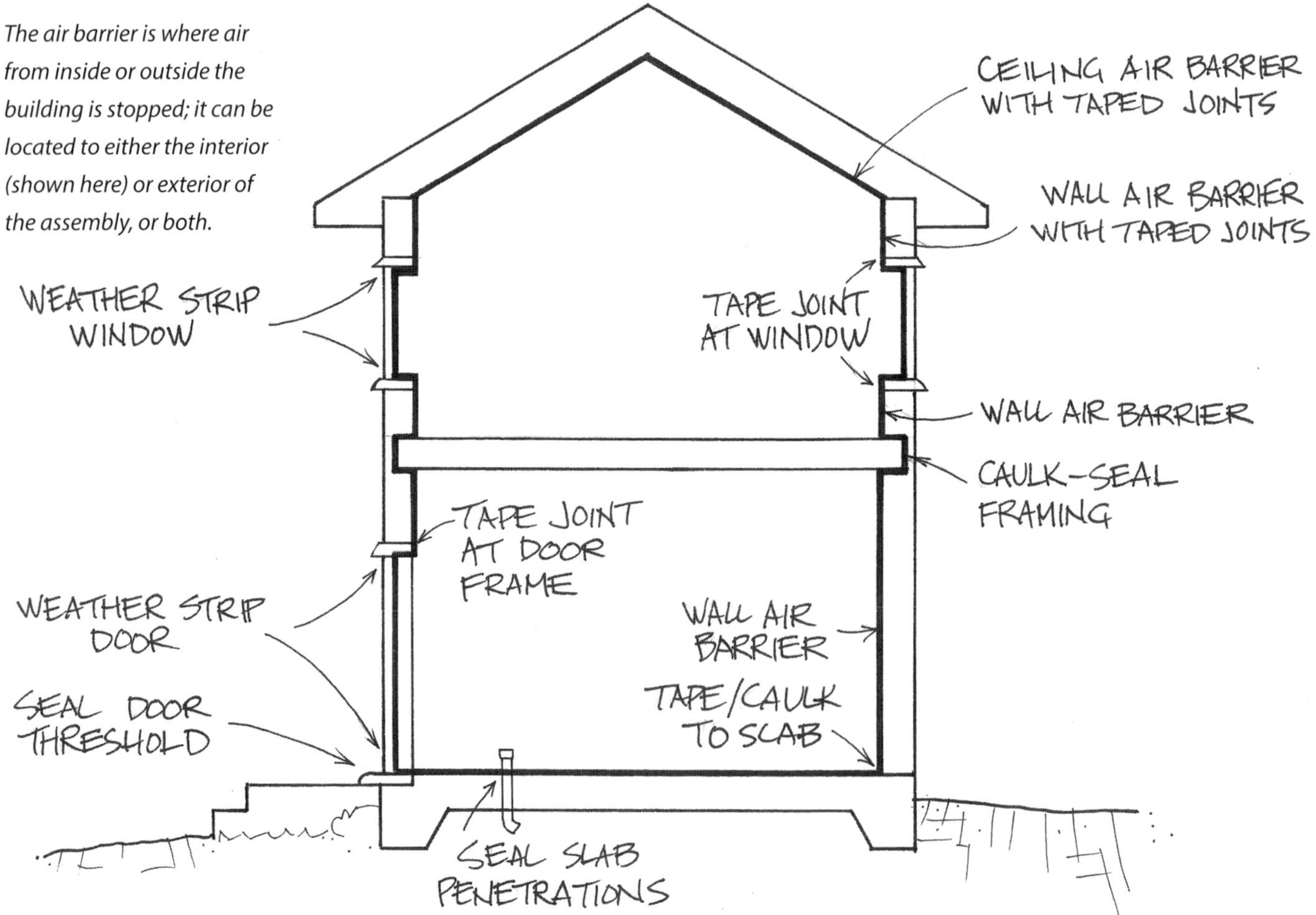

The air barrier is where air from inside or outside the building is stopped; it can be located to either the interior (shown here) or exterior of the assembly, or both.

discomfort from temperatures that are too high or low, as well as drafts that can further exaggerate discomfort through the experience of turbulence.

- **Degradation of thermal performance** — Loss of tempered air from leaks through the building enclosure has a considerable effect on fuel consumption during heating and cooling seasons. Decreases in efficiency from leakage can be more significant than those caused by inadequate thermal control.
- **Mold, mildew, and rot** — Air leakage through the building enclosure can carry significant amounts of moisture into the enclosure, and can lead to issues of mold and mildew growth and rot in structural and insulation materials.

Key Concepts for the Air Control Layer

In order to ensure that your air control layer functions adequately, there are a few important concepts to understand:

- **Airtightness is important** — There is a common misconception that it is possible to make a building "too airtight." The importance of airtightness for comfort, occupant health and safety, thermal performance, and building durability cannot be overstated: *No benefits come from intentionally allowing air leakage.* It is a very poor strategy to rely on building leakage to supply your home with "fresh" air. Active ventilation (see below) is an important component of airtightness strategies.
- **Airtightness should be planned** — Accurate energy modeling of your building will require a target for airtightness as a key variable. It is very informative to adjust the rate of air leakage in your energy model and witness the effect on energy performance. Lowering the rate of leakage can, in some circumstances, have a greater effect on efficiency than raising insulation quantities. Choosing a target for airtightness is central to planning for energy efficiency.
- **Airtightness should be tested** — Planning a target for airtightness is important, but it's worthless unless you ensure the target is reached in construction. Airtightness testing using a blower door should be carried out as soon as the air control layer is complete (and accessible) so you can fix any problems *before* the layer is covered by finish materials.
- **Airtightness requires active, balanced ventilation** — Healthy and efficient building assemblies require airtightness, while your home requires the exhaust of excess moisture, and the occupants require fresh air from a properly designed and installed ventilation system. System components will vary depending on climate conditions, but an appropriate strategy must be chosen and implemented.

  Active ventilation is often derided as an "iron lung" for the building, but if you are willing to have active systems for heating, cooling, refrigeration, hot water, laundry, and other services, then an active ventilation system — often consisting of a fan or two in a box — should not be viewed as contrary to your building objectives.

  "I'll just open a window" is often used as an excuse for not designing an active ventilation system. This can be part of a ventilation strategy, but it doesn't work well on still days, on days when interior and exterior temperatures are similar, or in rooms with a single window and a closed door. It's also extremely inefficient when the temperature differences between inside and out are high and does not work at all when not operated due to comfort or neglect.
- **Air leakage rates** — The flow of air through the building assembly is most commonly expressed in residences as *ACH50* — the number of times the volume of air in the building will change per hour when there is a 50 pascal

pressure difference between the interior and the exterior. A leaky older home may be 8–10 ACH50, while current code requirements are between 2–5 ACH50. Extremely efficient homes are less than 1 ACH50.

- **Air leakage, mold, and rot** — Air leaking through the building assembly can carry a lot of moisture, and as the leaking air cools down, it can deposit that moisture as condensation inside the assembly.

  The quantities of moisture can be appreciable: a leak in an exterior wall of just 1 square inch (6.5 cm^2) can move as much as 30 quarts (28.4 liters) of water into the assembly *over the course of a single heating season in a cold climate.*[16] This amount of water is enough to grow mold and instigate rot, especially if the assembly is unable to dry out.
- **Small leaks are important** — Airtightness requires diligence to be effective, as the size of a leak is not proportional to the amount of loss that will occur. An analogy is covering half the opening of a hose with your thumb: the amount of water flowing is not reduced by half, instead a similar amount of water comes out, just at a higher pressure. A half-hearted attempt to create an air control layer is not sufficient, and the details are critical.
- **Materials for air control layer** — Building enclosure materials used for the air control layer must have an air permeance rate no greater than 0.02 L/(s•m^2) at a pressure difference of 75 Pascals when tested in accordance with ASTM E 2178. Air barrier materials should meet the requirements of the CAN/ULC S741 Air Barrier Material Specification. Many common building materials can be used as an air control layer. Ideally, materials for the air control layer have as few joints and seams as possible, and are simple to seal at each seam.
- **Continuity** — The weakest zones for air control layers are transitions, joints, and seams between different materials. As air will move through tiny gaps and holes, a plan for ensuring continuity throughout the entire air control layer is critical.
- **Design to minimize penetrations** — Service cavity walls allow for plumbing and wiring to

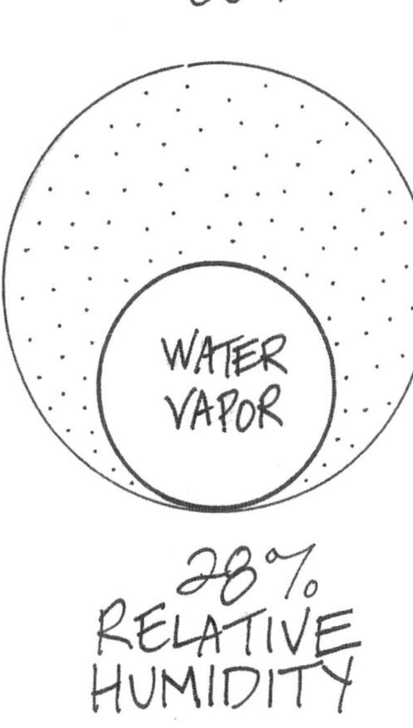

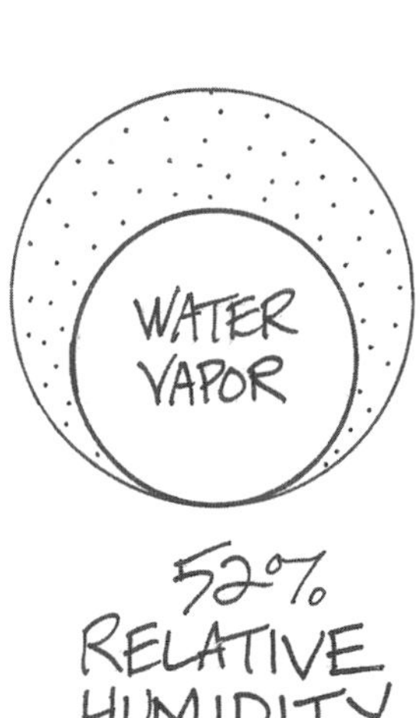

As air temperature drops, the air gets denser, and the relative humidity increases, even though no additional vapor has been introduced into the system.

stay on one side of the air barrier, greatly minimizing the number of penetrations.

- **Durability** — As very small punctures in the air control layer can have dramatic negative effects, the air control layer will ideally be durable over the lifespan of the building, and holes or punctures will be easy to see and repair. Thin, sheet-style barriers are often inherently weak and can be damaged during construction or in common use.

Putting It into Practice

At the design phase, the air control layer requires careful thought to ensure appropriate placement and continuity. There can be many interruptions to the layer, and to maintain continuity, you must have a strategy for each of them, including electrical boxes and wiring, plumbing and HVAC penetrations, door and window openings, and intersections with interior walls and exterior elements like porches. Wherever possible, design for service cavities that minimize the number of penetrations.

Feasibility during construction must also be considered, since a continuous line on a drawing may not be able to be connected if the order of operations during building will hide or bury a crucial connection.

When designing an airtight building, a whole-house ventilation and pressure balancing strategy must be used. Combustion appliances (wood stoves, fireplaces, gas heaters) must be direct-vented or supplied with make-up air, or else combustion gasses may leak or be drawn into the building.

At the construction phase, solid communication between the designer(s) and the builder(s) will ensure that everybody on the construction site understands the air control layer and its detailing.

The air control layer requires careful installation and monitoring to ensure continuity. The

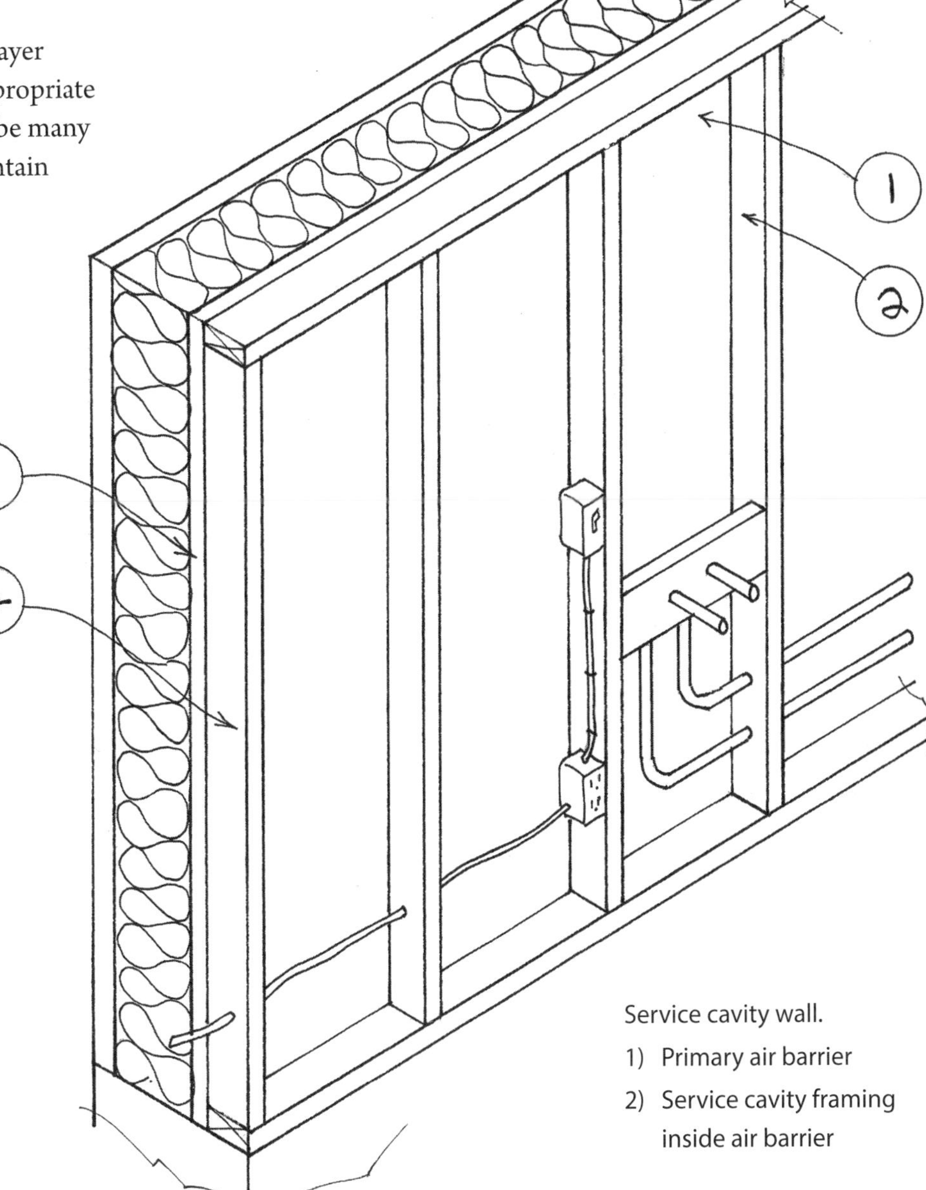

Service cavity wall.

1) Primary air barrier
2) Service cavity framing inside air barrier

A blower door is used to depressurize the building to test for airtightness.
CREDIT: CHRIS MAGWOOD

build team must be aware of its placement in the assembly and the means of connection between elements. Intentional punctures and penetrations must be made conscientiously, as each hole can contribute significantly to the performance of the air control layer.

As soon as the air control layer is in place, test the building using a blower door. While the blower depressurizes the building, you will learn in real time if you have met your airtightness target, and you will also be able to find, and hopefully seal, any leaks that are detected during the test. It may be prudent to run this test again prior to occupancy, to find and address any issues with the air control layer that may have occurred during the remainder of the construction process.

Your air control layer is a critical component of your project. It is a consideration that must be determined early in the design process and then followed through the entire construction process by a person designated to carry this responsibility to ensure that your goals for occupant health, energy efficiency, and building durability are met.

Vapor Control Layer

The vapor control layer is subject to a great deal of debate, and misinformation abounds about its role and function.

What Does the Vapor Control Layer Do?

The air inside and outside our buildings always contains water vapor in varying quantities. A *vapor drive* (caused by a gradient in humidity and temperature across the building enclosure) forces *diffusion,* the movement of water molecules through a solid material. Diffusion occurs without the movement of air through the building enclosure; it takes place through solid, airtight materials. The role of the vapor control layer is to *manage the diffusion of moisture into the building enclosure.*

You will notice that this definition does not insist that the vapor control layer must *eliminate* or *prevent* the diffusion of moisture into the building enclosure, but simply to *manage* it. Common building practice includes the use of a *vapor barrier* as the vapor control layer, in an attempt to prevent any vapor from diffusing into the building enclosure, but this is not always necessary, and may, in fact, create problems, as explained in this chapter.

Why Is the Vapor Control Layer Important?

- **Mold, mildew, and rot** — Excessive vapor diffusion into the building enclosure can lead to the accumulation of significant amounts of moisture, which can condense into water and result in mold and mildew growth and rot in structural and insulation materials.
- **Degradation of thermal performance** — Accumulation of moisture in the thermal control layer due to the condensation of excessive vapor diffusion can dramatically lower the performance of many types of insulation.

Key Concepts for the Vapor Control Layer

- **Vapor drive** — The rates and directions of vapor drive are based on several factors:
  - *Concentration gradient* — Movement from high to low absolute humidity; related to vapor pressure.
  - *Temperature gradient* — Movement from high to low temperature.
  - *Air pressure gradient* — Movement from high to low air pressure.

Combined, these factors create a *vapor drive,* forcing water vapor molecules to *diffuse* through building enclosure materials in the direction of the vapor drive.

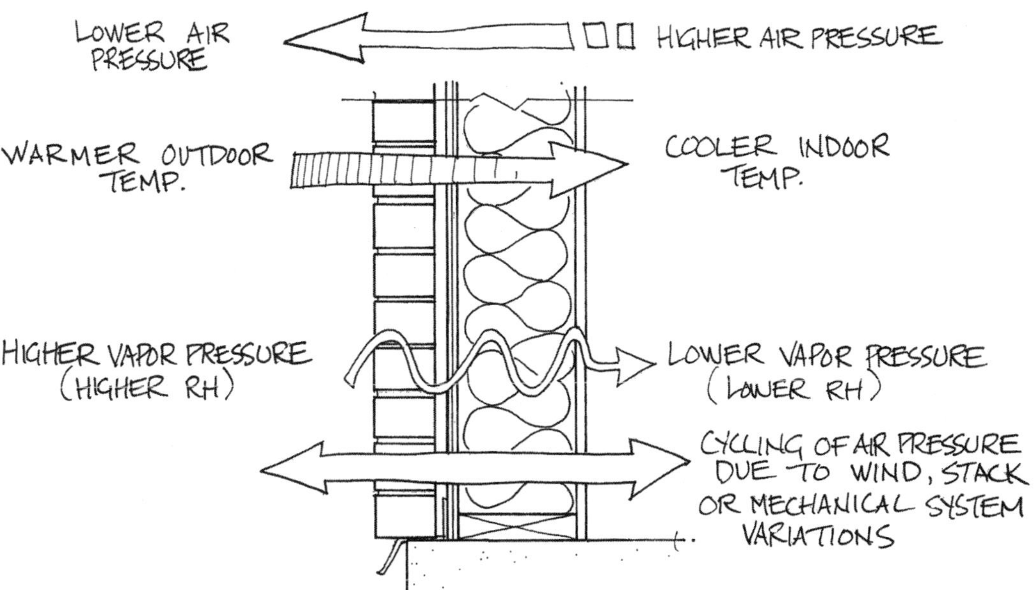

Vapor drive is governed by humidity, temperature, and air pressure. Accordingly, the vapor drive changes in both direction and intensity over the course of any given day. Often, relative humidity and temperature will induce a drive in one direction, overriding relative air pressure pushing in the opposing direction, as is pictured here (higher relative humidity and temperature outdoors, higher air pressure indoors).

- **Diffusion** — The movement of water through a material. Diffusion can occur in two main ways: vapor diffusion and liquid diffusion. The water control layer addresses liquid diffusion, and vapor diffusion is addressed by the vapor control layer.

It is important to understand that vapor diffusion can introduce moisture into a building assembly, and it can also allow moisture to exit the assembly; diffusion is both a wetting force and a drying force depending on the conditions of the vapor drive. In fact, when we refer to walls or roofs being able to "dry out," we mean the mechanism of liquid water evaporating into vapor and diffusing through permeable materials, either to the inside or outside.

- **Perms** — Different materials allow diffusion due to vapor drive at different rates, and the ability of materials to retard diffusion is measured in *perms*. There are four classes of permeability, as seen in Table 6.1.

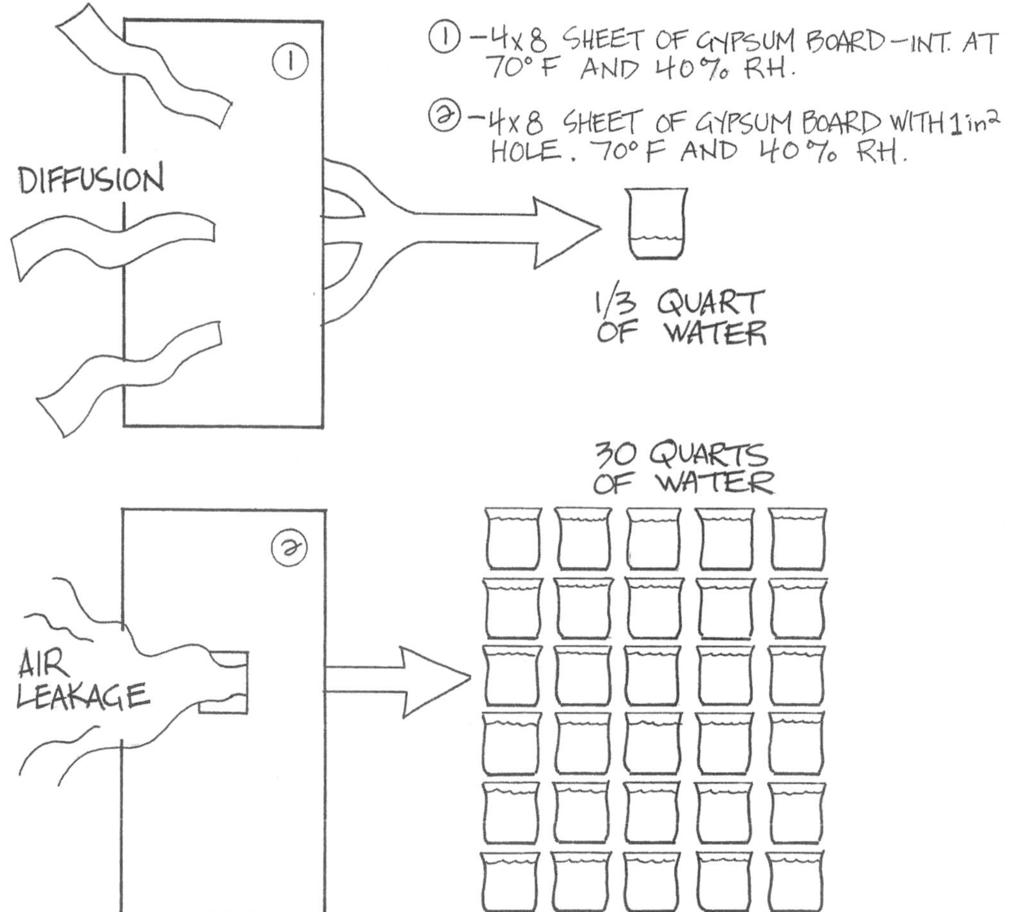

While vapor-permeable materials have the potential to allow water vapor into an assembly through diffusion when the vapor drive is sufficient, far greater concentrations of water vapor can be carried into the assembly through air infiltration; shown in this illustration is the aggregate amount of moisture that moves through these two mechanisms over the course of an average heating season in a cold climate. (Source: Lstiburek, Joe: "Insulations, Sheathings, and Vapor Retarders," Research Report 0412, 11/04, buildingscience.com)

Some common enclosure materials are shown in this chart. You should know the class of permeability of all materials in your building enclosure, and they must be examined for their collective effect on vapor diffusion in both directions.

- **Vapor barriers** — Materials that are Class I vapor retarders (typically called *vapor barriers*) can be used as the vapor control layer. A vapor barrier will prevent diffusion into an assembly, but will also prevent drying by diffusion out of the assembly.
- **Vapor permeable assemblies** — Assemblies that are composed of vapor permeable materials will allow diffusion from either side of the assembly and will also allow drying from either side. For permeable assemblies, the air control layer will function as the vapor control layer.
- **Vapor throttle** — Assemblies can incorporate a vapor control layer of Class II or III vapor retarders to reduce, but not prevent, the diffusion of moisture. This vapor control layer would be employed on the side of the wall with the dominant vapor drive. This will still allow for drying potential out of the assembly when humidity levels build up sufficiently within the assembly.
- **Moisture balance** — The vapor control layer is intended to ensure that diffusion does not cause moisture problems in the building enclosure. But how much moisture is too much? This depends on maintaining a moisture balance of diffusion, storage, and drying.

As this illustration shows, a moisture balance is maintained when wetting (due to water leakage, air leakage, and diffusion) is able to dry due to diffusion

Table 6.1

VR Class	Description	Permeability	Materials
Class I	Vapor-impermeable (vapor barrier)	≤ 0.1 perm	Sheet polyethylene, sheet metal, aluminum foil, foil-faced insulation, rubber membranes, glass
Class II	Vapor-semi-impermeable	> 0.1 to 1 perm	Kraft-faced fiberglass batts, oil-based paint, vinyl wall coverings, XPS foam > 1" thick, cement stucco
Class III	Vapor-semi-permeable	1 to 10 perm	Plywood, OSB, EPS foam, fiber-faced polyiso foam, most latex paint
Vapor-open	Vapor-permeable	> 10 perm	Gypsum wallboard, earthen and lime plaster, asphalt-impregnated felt, most housewraps

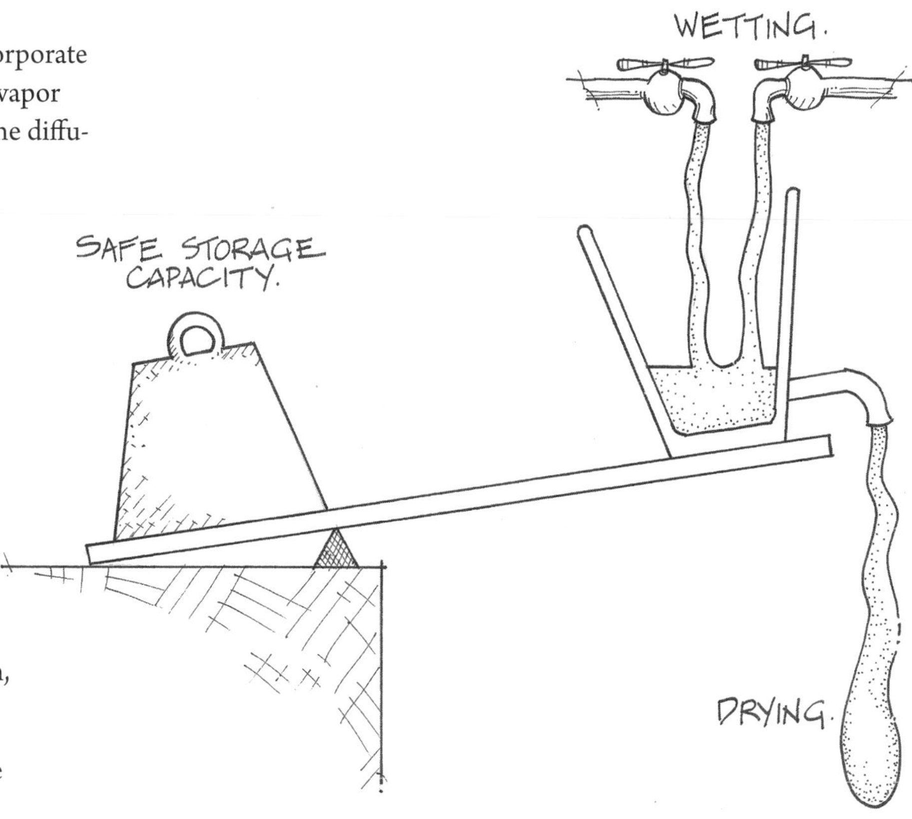

A balance of storage, drying, and drainage can be designed to allow assemblies to manage significant wetting without deterioration. The key word here is designed!

without exceeding the safe storage capacity in the assembly materials.

These three variables — wetting, drying, and storage — vary depending on climate conditions and material choices. A vapor control layer must be designed to meet the particular needs of your building in your climate.

- **Hygrothermal modeling** — Software exists to model the movement — and potential accumulation — of moisture and heat through an assembly over a period of time in particular climatic conditions. This type of modeling can help to ensure that the vapor control layer will function adequately for high-performance assemblies in cold or hot-humid climates, or where innovative or experimental materials or systems are being used. Hygrothermal modeling is more complicated than energy modeling, and should be conducted by an experienced professional.

Putting It into Practice

At the design phase, it is important to develop a vapor control strategy that is based on climatic conditions and material choices that meet the goals set for all of the control layers. This may involve a combination of vapor control strategies for different parts of the building enclosure.

Airtightness is of crucial importance for vapor control, as air leakage will move much larger quantities of moisture into the building enclosure than diffusion. Diffusion becomes more important once a building is relatively airtight, but the first concept in vapor control is air control.

Building codes in many jurisdictions use the language of vapor barriers, rather than vapor retarders, and the use of a strategy other than a vapor barrier may require the use of alternative compliance pathways.

At the construction phase, it is important that materials specified for the vapor control layer in the plans are actually used on site. Competing materials that may seem to be equivalent from a construction perspective (for example, different brands of house wrap, sheathing boards, and paint, among others) can have very different permeability ratings, and the substitution of a product that does not meet permeability requirements will negatively affect the performance of the vapor control layer and may result in moisture damage in the building assembly.

Some materials used as a vapor control layer must be installed with the proper side of the material facing the interior in order to perform as intended.

Helpful Analogies for Building Control Layers

It can help to summarize the different control layers of a building if we think about them like layers of clothing we can put on our bodies:

- ***Wearing a sweater is like putting on a thermal control layer.*** Wool is a good insulator, slowing the flow of heat from our bodies, and the thicker the sweater, the warmer we will be. But when the wind starts to blow, the sweater is not very effective. And if it starts to rain, the sweater does not keep us dry (though it will dry out after a soaking, if the conditions are right).
- ***Wearing a rain suit is like putting on a weather resistant barrier.*** A plastic layer can shed the rain and keep us dry, and it will also keep the wind from our skin. But it's not effective for keeping the heat inside.
- ***The rain suit is also like putting on a vapor barrier.*** We soon find that moisture from our bodies gets trapped inside the impermeable plastic and starts to condense on the inner surface, causing us to be wet and uncomfortable.
- ***Wearing a Gore-Tex jacket is like putting on a vapor permeable air control layer and WRB.*** It will repel most of the rain, and it will keep

out the wind, helping to keep us comfortable. It will also allow moisture from our bodies to transpire through tiny pores in the fabric, helping to keep us dry on the inside. But it will not keep us warm.

- ***Wearing a Gore-Tex jacket over a sweater is great for long-term comfort.*** Rain and wind are repelled, moisture is allowed to transpire through the sweater (which can store a lot of moisture away from our skin) and then out through the Gore-Tex. In the still air under the Gore-Tex, the thermal control of the sweater keeps us nice and warm.

This clothing analogy can work to clarify some of the basic principles of weather protection, airtightness, and thermal insulation. It also points to the importance of climate — the outfit would be very different in an equatorial rainforest than a temperate-zone winter or a hot desert. In many parts of North America, conditions are highly variable, which means our buildings must be "dressed" for the worst weather extremes yet still function in more moderate conditions. Imagine yourself dressed for winter but standing outside in summer. Shade and reflective colors will help you stay cool, and well-placed zippers (windows) will allow cooling airflows to reach your skin.

Understanding this analogy does not make one a building scientist, but imagining appropriate clothing strategies can be very helpful to work through a basic understanding of performance and comfort in buildings.

Thermal Mass

The subject of thermal mass can be confusing, as the effect of thermal mass is often equated with thermal insulation. However, *thermal mass is not insulation.* Insulation limits or restricts the flow of heat; thermal mass freely and easily absorbs and releases heat. The confusion between thermal mass and insulation can arise because, under certain conditions, the *thermal performance* of a massive material can lend itself well to occupant comfort.

Consider the two scenarios pictured on page 82. The empty tent has almost no thermal mass, being constructed of lightweight material and containing nothing but air (and an occupant). The tent containing the boulder has a huge amount of thermal mass.

The empty tent will quickly cool down on a cold night. Then, when the sun hits the tent in the morning, it will quickly get warmer. The lack of thermal mass means that the interior temperature can change very quickly because only the air in the tent must change temperature.

The tent with the boulder will remain warmer longer, because the heat stored in the mass of the rock will take some time to be lost to the night air. Then, when the sun hits the tent in the morning, it will take longer to warm up because the mass of the boulder will absorb a lot of thermal energy before its surface temperature rises.

If both "tents" offer the same degree of thermal insulation and airtightness, neither scenario is inherently better than the other. The tent with the boulder requires a lot of thermal energy to raise its temperature, and then releases a lot of thermal energy as it cools. The empty tent requires a very small amount of thermal energy to change its temperature, and then cools quickly.

Now translate these scenarios to a building. You wake up in the morning and turn on a heating device. In a building with little thermal

mass, the temperature will reach the desired level relatively quickly. In a building with a lot of thermal mass, the desired temperature will take much longer to reach. The low-mass building will require more frequent but shorter inputs of heating energy and the temperature swings will be shorter. The massive building will require fewer but longer inputs of heating energy and the temperature swings will be longer.

HEATS UP FAST...
...COOLS DOWN FAST
...BOULDER TAKES A LONG TIME TO HEAT UP AND COOL DOWN...

The ideal amount of thermal mass in a building is dependent on the climatic conditions, the type of heating system being used, the amount of passive solar exposure, the amount of insulation, and occupancy patterns (do occupants want faster or slower response times from heating and cooling systems?). More or less thermal mass is not necessarily better, but should be among the important design considerations for the building.

Passive Solar

The sun is the source of all energy on our planet, and as such it seems foolish to ignore the impact of the sun on our buildings. Passive solar design encourages us to think about how the sun will relate to our building, and asks us to plan how we can best *exclude* direct solar heat gains during cooling seasons and *include* direct solar heat gains during heating seasons.

By doing so, we can realize significant reductions in the amount of energy we need to condition our spaces.

There are two aspects to solar exposure that must be considered:

- *Bearing angle* — The point on the horizon where the sun appears in the morning and disappears at night.
- *Altitude angle* — The height of the sun above the horizon.

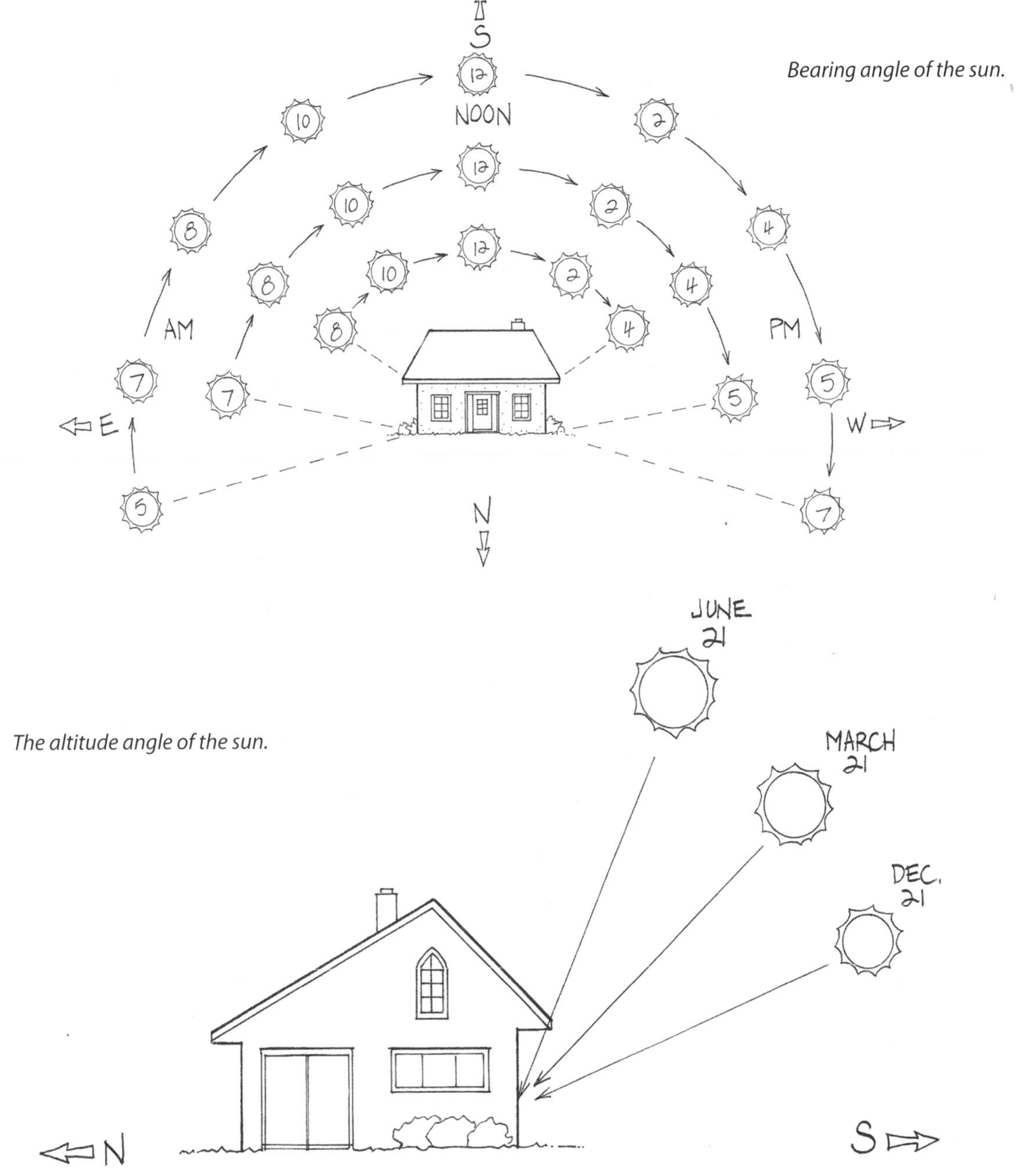

Bearing angle of the sun.

The altitude angle of the sun.

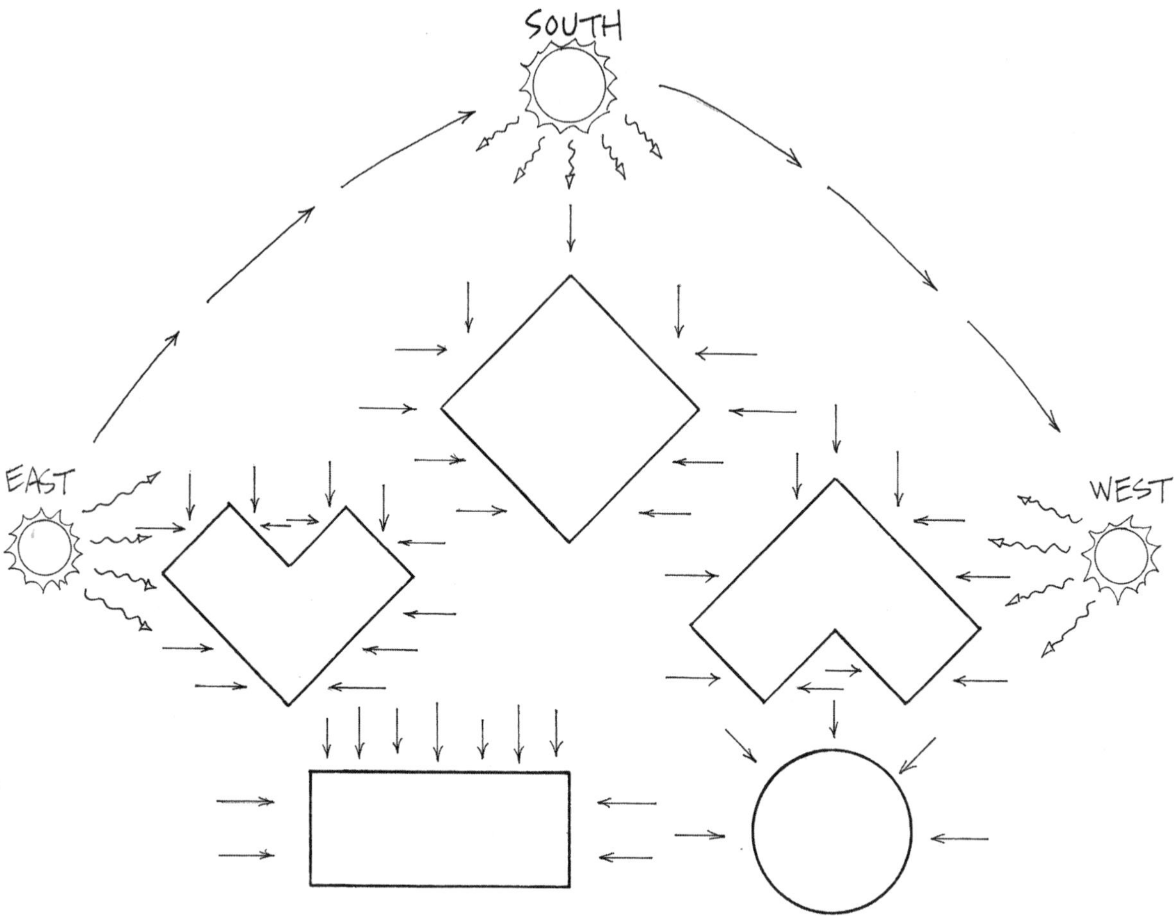

Many different building shapes can be oriented to take advantage of passive solar gains.

Depending on the latitude of your building site, these points will change — sometimes dramatically — from summer to winter. Understanding these angles allows you to strategize for the seasonal inclusion and exclusion of direct solar gain.

Many building shapes and orientations can accommodate passive solar design. There exists the notion that "best practice" involves creating a building with a long east-west axis that maximizes southern exposure. However, as the drawing above makes clear, a building of any shape can receive a lot of solar exposure. Regardless of whether or not solar exposure is ideal, it is worthwhile to incorporate passive solar planning, which includes gain and loss calculations, but also room layout, heat distribution, and the mass, thermal capacity, and reflectivity of building materials. It is better to make the best use of whatever amount of solar exposure and/or to take advantage of whatever exclusion aspects are available on a given site. *There is no site and no building design that cannot benefit from a passive solar analysis.*

Roof overhangs are the simplest form of passive solar design. As shown in the examples above, the roof overhangs are designed to allow for maximum solar gain in the windows in the winter and maximum solar exclusion in the summer. Despite the narrow southern exposure on this particular house, 25% of the annual heating energy requirements come from free sunshine, and air conditioning is rarely needed. The energy impacts are real!

Arkin Tilt Architects
Sage Design Build
Eastern Sierra Residence
Gardnerville, Nevada
2005

Please refer to Table 5.1, pages 58–59 as reference for the icons used in this section.

Situated on the eastern slope of the Sierras overlooking Carson Valley, this home works with slope and orientation to celebrate the rugged beauty of its site while sustaining against its harsh climate. Alternative construction techniques, solar technologies, salvaged materials, and daylight are integrated to create a dynamic, efficient, and livable home. PHOTO CREDIT: ED CALDWELL

Parallax R:E:D
Tapial Homes
Mark Allen residence
Huntsville, Ontario
2012

The main objective for this build was to create a comfortable and healthy home to raise a family in. Utilizing the Passive House approach to design, energy modeling, and airtightness, this home has outstanding resiliency. It features R-40, 18″-thick rammed earth walls on the exterior, and a custom masonry heater as its primary heat source. The 7.5 kW roof-mounted solar array offsets the family's entire annual energy usage, meeting Net Zero targets.

EXTERIOR PHOTO CREDIT: EVER IMAGES, INTERIOR PHOTO CREDIT: TAPIAL HOMES

Chris Magwood/Dale Brownson
Sandra Zabludofsky residence
Madoc, Ontario
2004

4 4 3 4 4 4 3 4 2 2

This 1,300 sq. ft. home was designed and built to embody the "100-mile house" ethic. The vast majority of materials were produced or manufactured in the immediate area. Indoor environmental quality was a high priority for this off-grid home, which was the first in Ontario to feature exterior and interior earthen plasters. The home uses a hybrid pier-and-rubble trench foundation, a structural frame made from locally harvested wood, and hemp bale walls.

PHOTO CREDIT: DANIEL EARLE

New Frameworks
Ellison/Lannes Unicorn Farm
Middlesex, Vermont
2014

This off-grid high-performance home features an airtight assembly, highly insulated double-stud dense-pack cellulose walls, passive solar design, and reclaimed finishes and fixtures. The home is heated by radiant tubing supplied by a Froling pellet boiler and woodstove heat. The electricity is supplied by solar PV. Almost all of the wood was locally sourced, and all materials were selected with an eye on lowering the embodied carbon of the structure.

PHOTO CREDIT: NEW FRAMEWORKS

4 4 4 4 3 4 4 3 4 4

Frank Tettemer Design
Living Sol Building & Design
John and Lynn Epps residence
Eganville, Ontario
2015

3 2 4 3 2 3 4 3 3 4

This home has a footprint of 1,800 sq. ft. It has double stud wall construction 12″ thick with dense-packed cellulose, and 24″ of blown-in-place cellulose in the attic, as well as R-30 basement walls and floors to assure that the stone masonry heater will only need a single morning firing per day. Standard HRV was replaced with an air handler to take outdoor air and hot air above the masonry heater, and distribute it to all parts of the home. Electricity is stored downstairs in a Surrette Big Red battery bank, and distributed through the house and garage/workshop, using dual Magnum inverters. Solar photovoltaic panels are ground mounted; the array produces 4,500 watts. Photo credit: Lynne Epps

Jeff Waring, with Harvest Homes as consultant
Evolve Builders Group and Harvest Homes
Horseshoe Valley House
Orillia, Ontario
2009

3 1 3 3 3 3 3 3 3 4

This high-efficiency estate home was designed to integrate natural and found materials including adobe floors, earthen architectural reliefs/sculptures, local stone, timber framing, and straw bale walls while also providing comfort and resiliency against power outages and road closures through means including rainwater harvesting, a wood-burning high efficiency masonry heater, and both solar photovoltaics and solar hot water collection.
Photo credit: Evolve Builders Group

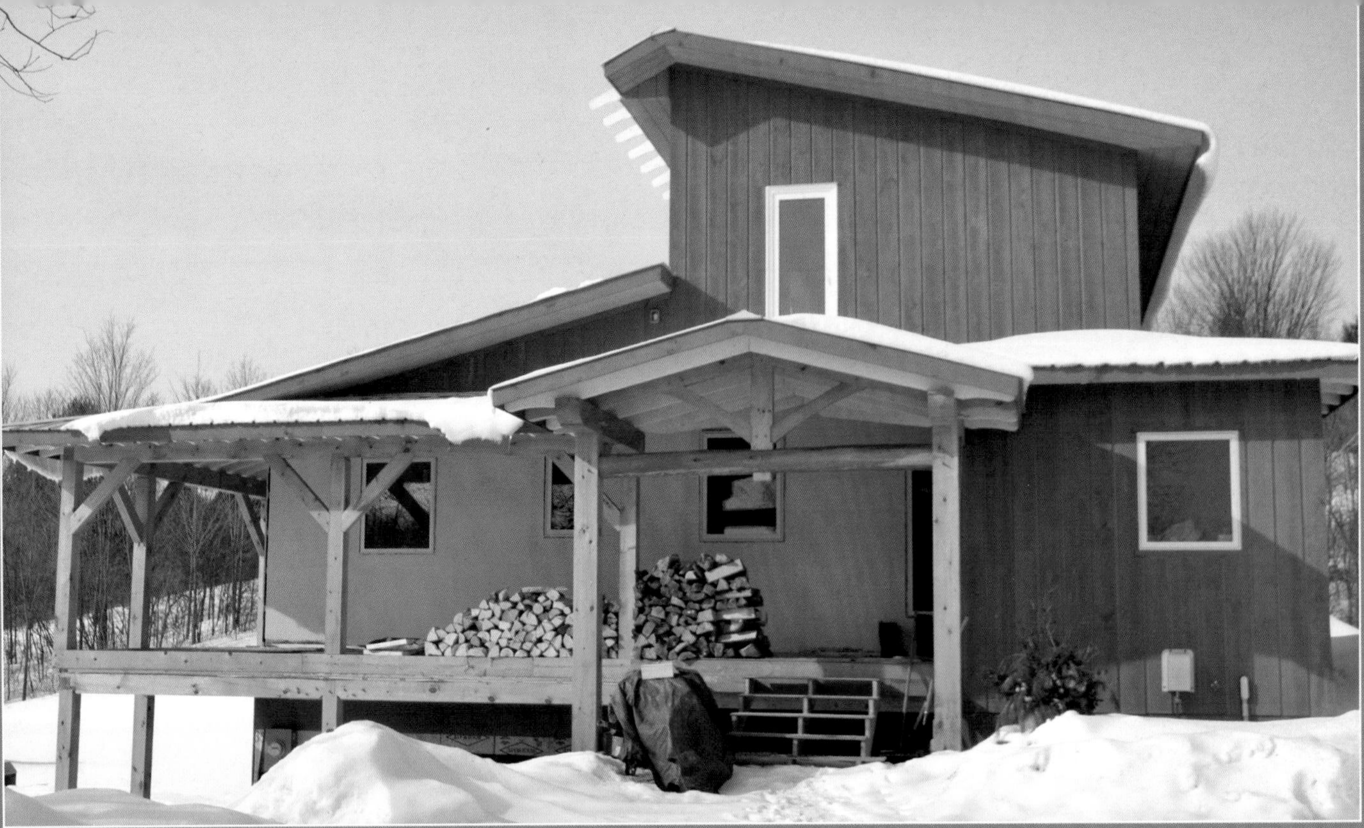

Dwellings Design/Build
Ratzinger residence
Maberly, Ontario
2013

This house is part residence and part chocolate factory for Fine Chocolate by Ludwig. *This is a 3,000 sq. ft. timber frame house built using a prefabricated straw panel system; it has hydronic heating throughout powered by an interior wood boiler and solar hydronic panels.* Photo credit: K. Bailey

Endeavour Centre
Canada's Greenest Home
Peterborough, Ontario
2013

This urban infill project in central Peterborough, Ontario, was Endeavour Centre's first attempt to design a "zero house," with goals of achieving a zero carbon footprint, zero fossil fuel use, zero energy bills, and zero toxins. The home brings natural building materials into an urban context. A Durisol foundation, prefabricated straw bale wall panels, and a focus on local materials are among its features.
Photo credit: Chris Magwood

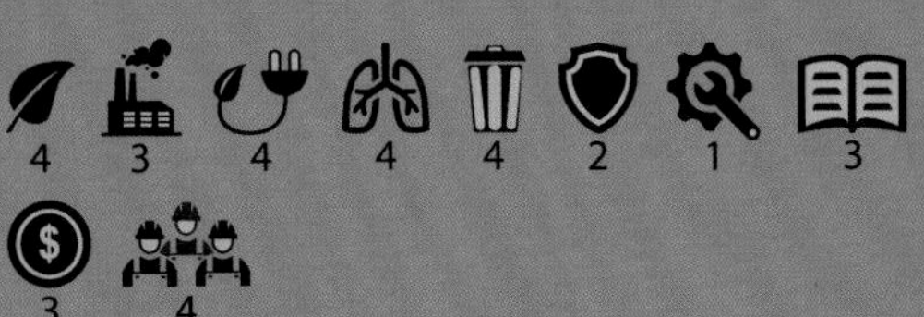

Evolve Builders Group & Harvest Homes
Carter House/Ben Polley & Jennifer Woodside
Guelph, Ontario
2009

A century-old, designated heritage home, "Carter House" was originally constructed by the co-founder of the Canadian Cooperatives Association. A renovation and a straw bale addition were designed to preserve the building for the next century. Energy and water conservation, and occupant resilence were chief among the goals, though always only with materials and methods consistent with or borrowing from historical aesthetic expectations.
PHOTO CREDIT: EVOLVE BUILDERS GROUP

Arkin Tilt Architects
MS Builders
Santa Cruz Strawbale
Santa Cruz, California
2010

Avid surfers and professors of Biology and Environmental Studies in Santa Cruz, CA, the owners wanted to push the ecological envelope while providing a fun, comfortable house for their family of six. The result is a 2,500 sq. ft. two-story house that includes four bedrooms and an office, as well as a small (330 sq. ft.) one-bedroom accessory dwelling unit with its own entrance. PHOTO CREDIT: ED CALDWELL

Endeavour Centre
Straworks Inc.
Armstrong residence
Cobourg, Ontario
2015

This is the builders' largest and most involved project, but also their most natural and locally sourced. The main floor exterior walls are insulated with straw bales from a neighboring farm and plastered with lime and local clay. Pine and cedar trees from the property were felled, milled, and used for the exterior porch's cedar shingles. The main floor's compressed earth blocks were made just a few miles away by Fifth Wind, using primarily local clay and sand. Two-thirds of the house sits on a rubble trench foundation made of local aggregate. The house is plumbed for greywater reuse and is heated by a pellet boiler; heat is distributed under 3,800 sq. ft. of floor over the two stories.

EXTERIOR PHOTO CREDIT: CHRIS MAGWOOD, INTERIOR PHOTO CREDIT: DEIRDRE MCGAHERN

Agostino Terziano Design
Tapial Homes
Mary McCulley residence
Huntsville, Ontario
2016

The program for this build was to create a small, affordable, energy-efficient home located in town, close to all amenities. This home features 20″-thick rammed earth walls, a frost-protected shallow foundation, triple-pane fiberglass windows, and R-80 attic insulation. A small, energy-efficient wood stove and in-floor electric heat combined with an HRV provide a high level of comfort in the home year round. PHOTO CREDIT: EMILY BLACKMAN

Arkin Tilt Architects
Earthtone Construction
Healdsburg Family Residence
Healdsburg, California
2015

Nestled in a beautiful 170-acre property along Porter Creek in rural Sonoma County, the design and construction of this off-grid homestead create a harmonious fit with minimal impact on both the specific and global landscape, providing a sustainable frame for living and raising twin daughters. The indoor/outdoor flow creates an enriching yet peaceful living space for the family and their deep appreciation of the natural world. Photo credit: Ed Caldwell

Frank Tettemer Design
Living Sol Building & Design
John and Denise Steinman residence
Barry's Bay, Ontario
2016

An energy-efficient four-season cottage designed to fit the prescribed minimal footprint of 840 sq. ft. The home features local cedar, spruce, and white pine, with landscaping basics for the owners to plant using zeroscape permaculture. There is an adjacent storage shed/guest bunkie, an attached screen porch, and a second-floor bedroom balcony overlooking Trout Lake. All exterior pine and cedar timberframe materials were left untreated — to age and grey naturally. Pine/cedar chips were used as initial ground cover; interlock bricks used as patio stones allow drainage; and local stone steps lead to a lakeside dock. Photo credit: New Frameworks

Nicolas Koff, Office O-U and Evolve Builders, as consultant
Evolve Builders Group & Harvest Homes
K-House
Ancaster, Ontario
2015

This home provides a look that is unforgivingly modern while borrowing heavily from the traditional 2+1 story massing common to the neighborhood; materials were specifically selected to provide good indoor air quality and require minimal maintenance. The house combines state-of-the-art technology, such as an air-source heat pump and a direct vented wood-burning fireplace, with natural materials, such as oil-finished wood flooring, silicate paint on the walls (inside and out) and shou sugi ban *(charred) locally harvested eastern white cedar.* PHOTO CREDIT: NICOLAS KOFF

Andy Mueller — Greenspace Collaborative
Dwellings Design/Build
Weil residence
Westport, Ontario
2014

2 2 3 3 3 2 2 3 3 4

This passive solar off-grid retirement home features straw bale walls with earthen and lime plasters, some timber-framed elements, and reclaimed doors; its cherry sills and other accents were made from a cherry tree that came down on the owner's property.
PHOTO CREDIT: K. BAILEY

The roof overhangs of this home completely block out the sun in the summer (left), and allow full sun access in the winter (right).
CREDIT: CHRIS MAGWOOD

Incorporating Good Building Science into Your Project

Basic building science principles are relatively easy to understand and can offer a straightforward and "common-sense" approach that can be applied as you see fit for your project. It all boils down to asking a short list of questions every time you consider a design or construction detail:

- How is water excluded and shed in this part of the building?
- How is air control (airtightness) achieved in this part of the building?
- In which direction might vapor drive, and how is vapor controlled in this part of the building?
- How is thermal control (insulation) provided, and is it adequate in this part of the building?
- What heat flow(s) exist, and do they create or disrupt occupant comfort?
- How does the sun interact with this part of the building in each season?
- Does the thermal mass of the building suit the climate and the heating strategy?

These questions are relatively simple to consider for many of the larger elements of the building, but become more complex at every joint, seam, and intersection where different materials in the assembly meet. A good building design has a good answer to each of these questions at each junction.

These building science principles are relatively new to the building industry and certainly not built into the "common sense" of all designers and builders. In selecting your project team members (see Chapter 8), be sure to find out the degree to which they are familiar with at least the basics of building science, especially in terms of practical execution. If they are not asking themselves the questions listed above, then your building will not be as good as you should want it to be.

Good designers and builders will understand these principles at a deeper level and will often use modeling software and testing equipment to work through the complexities of incorporating them in the best possible way for your project. If the members of your team are unfamiliar with building science, computer modeling, or enclosure testing, you may want to ensure that at least one person with the relevant knowledge and experience is hired on to bring their expertise.

In the end, the application of building science to your project is about ensuring your own comfort, health, and safety and creating the conditions for your building to be sound and long-lasting — goals that are well worth achieving using all the tools building science can offer to help.

Chapter 7

The Design Game

EVERYTHING IN THE PREVIOUS CHAPTERS of this book is intended to help you clarify your overall goals for your project. But the goals outlined in your Criteria Matrix will need to migrate into a specific design. It's time to start playing the "Design Game."

Home design is a skill that many people train for and practice for years, and it may seem belittling to call it a game, given the amount of knowledge and experience that informs good design. However, approaching the initial stages of your design with a sense of playfulness can relieve the "performance anxiety" that often accompanies building design. If you embark on an adventurous investigation of possibilities, you will accumulate everything you need to know to inform a good design, whether you hire a professional to help you or complete the design yourself.

Don't leap into drawing plans

We typically associate home *design* with home *plans,* but they aren't the same thing. Refrain from immediately putting your ideas into a floor plan drawing; instead use the Design Game as an information-collection exercise. Once a drawing is put on paper, it is all too easy to become locked into the box you've drawn. Give yourself some time to consider broader questions first — answering these will make it easier to draw plans when the time comes.

What do you really want?

The first — and most important — step in the design process is to understand what you need and want from the design. Ask yourself the questions plainly, and take enough time to list, compile, and sort all the answers. There is no point too large or too small to be excluded from this list, nothing too obvious or too obscure. Build on this list — daily, if possible — by taking notice of the built environment around you: consider your current home, the homes of friends, things you see as you move about and travel, elements of public and work spaces that you appreciate. From the shapes and volumes of spaces to little details like knobs and handles, these are all pieces that can inform the more formal design process to come.

The Internet Era has introduced a whole new element to the Design Game, allowing you to find and archive photos from a nearly infinite source. While accumulating images, be sure to take note of what it is you find appealing in each photo: What is the image saying to you? Is it a specific element, a feeling it creates, the colors? A library of images is much more helpful if you understand what it was that first appealed to you. We now have instant access to the world's rich architectural history. Enjoy it and revel in it!

Sooner or later, you are likely to see some patterns emerge in the things that attract you. When your explorations start to deliver variations on similar themes, it's probably time to move on to the next steps.

Mapping your movements

Take the time to study and think about human traffic circulation in your home (if you have kids, it may be more like air traffic control!). Think about all the various kinds of movement: in and out of the house, between rooms, within rooms. Are these patterns different depending on time of day and time of year? If you entertain, notice what happens when extra people are in the space. What works well, what needs improvement? Take note of the same movement types in other homes and spaces. Add all of these observations — both positive and negative — to the notes you are compiling for your design. Your design will give you the chance to blend the best aspects of all you notice — and avoid the worst.

Include your family

A home design will affect everyone in your family, so be sure to involve everyone in the process. You'll be amazed at how attentive to detail kids can be if they are included in this game. Encourage everyone to think about his or her current wants and needs, and to imagine the future, too. Will more children be arriving? Will parents, siblings, or friends be visiting regularly or moving in? Can spaces be flexible and serve multiple purposes, or be easily adaptable to new uses?

Inevitably, conflicts between various needs will arise. Let them present themselves; there's no need to impose limits yet. During the Design Game, be open to all possibilities, and treat every option as though it were viable.

It's not a matter of life and death!

It is easy to become so wrapped up in the Design Game that you feel you have to get everything perfect. Don't sweat it — every house is full of compromises. Even warm-spirited, inviting, and comfortable homes — the ones you've been admiring in your research — have their share of problem areas and compromises. You will figure these out. By the time you've played your way through the Design Game, you'll have learned an awful lot about what you want your house to be. This awareness — even when it's an awareness of flaws and concerns — is invaluable.

Building a paper house

The ideas you've gathered during the first part of the Design Game will eventually be ready to be turned into drawings. It can be a valuable exercise to attempt your own initial drawings, even if you fully intend to hire a professional to do this work on your behalf. Translating your design ideas into renderings — even just roughly sketched — is a very instructive process. At the very least, it will help to create an understanding of the work a design professional will be doing on your behalf, but it may also provide you with a deeper and more direct connection with the whole design process.

Designs before plans

The terms *design* and *plans* are often used interchangeably, but they refer to different kinds of drawings:

- **Design** is about spatial concept.
- **Plans** are the technical drawings that make the execution of a design possible.

There is usually a gradual morphing of a design into plans. At the early stages, design drawings may only define the building's overall shape and size, and will likely rely more on proportion than actual dimensions. As details begin to be attributed, the drawings become more and more precise, until they are fully scaled and include all the relevant dimensions.

Give yourself time to design

There is a temptation to leap directly into creating plans, often by starting with the floor plan for the home. It is often beneficial to hold off on this aspect and allow yourself time to consider your home without drawing an outline around the perimeter and then attempting to fill it in with rooms. Spend more time concentrating on visualizing three-dimensional space as much as possible. Even if your drawing skills are limited, if you allow your imagination to form visions of the home that include volume and depth, you may find that you build a much stronger sense of the space. This will be very helpful when you begin to work with the two-dimensional drawings needed for the plan set.

The kind of spatial design being proposed here involves conceiving of the building from the *inside out,* rather than from the exterior walls *inward.* Imagine yourself taking a walking tour of the home you envision: What do you see when you open the front door? Walk forward a few steps, describe what you see. Turn left and right. Look up. Look down. Move through the space in this manner and record the results as best you can. Perhaps this is via drawings, or perhaps it is achieved verbally. Maybe it's best represented by a series of images you've collected. Remember, you can't be wrong! Even if the drawings or descriptions don't do justice to your vision, if they are sufficient to remind you of that vision, then they are doing their job.

Disposable drawings

Design work can be fun, and it can be frustrating. Draw and re-draw. If something isn't working, change the perspective and work from a different one. Don't be afraid to take a break — home design can be a long process, so allow yourself time to create something that you are truly happy with. Think of your drawings as disposable (but don't actually throw them out), and start over whenever you feel lost or unable to complete something. Label and date each drawing and keep them in a file together. If you revisit them every so often, you might surprise yourself with the degree to which you agree with yourself!

Paper versus computer

The advent of accessible design software for computers is both a blessing and a curse for the inexperienced home designer. While drafting programs can help you to turn out very professional-looking renderings, they can also limit your vision by creating solid definitions of the space too early in the process. Computers make very accurate rectangles and boxes, but they often constrain "visions" of space in these early phases. It will be a matter of personal preference whether you choose to work on your design on paper or electronically (or maybe both).

Add detail gradually

Your first drawings don't have to be accurate or beautiful. As you create a design that meets your needs and pleases you, you can gradually add detail. Begin to add depth/thickness to walls, and begin to draw doorways and windows to scale. Consider the direction in which doors will swing. Add permanent fixtures like sinks, baths, showers, and toilets; place counters, closets, beds, desks, and tables. Think about appliances. For two-story designs, begin to place stairs and landings. And even if you don't know the specific details, consider the location of heating devices and mechanical systems.

Dimensions

As you start to work with more accurate dimensions, you may want to carry a tape measure with you wherever you go. By measuring existing rooms and spaces, you can build a realistic understanding of what dimensions will work best for your design. Most people tend to overestimate the amount of space they will need. Only by knowing how your numbers translate into real-world space can you avoid over- or under-sizing. Consider how space will be used, not just the amount of space that's available.

Design for efficiency

It is not too difficult to design individual rooms to suit your needs. It is a real skill, however, to be able to join all the rooms efficiently. The least efficient spaces in a home are hallways. These take up a lot of square footage and often don't serve any purpose other than short-term foot traffic to connecting rooms. Where possible, arrange rooms to limit the number and size of hallways. If a hallway is unavoidable, try to have it serve additional purposes (storage, art display, outdoor view, reading nooks, for example).

Efficient design also accounts for the primacy of daily "chores" such as cooking, cleaning, laundry, and moving goods into the home and waste out of the home. Give these activities careful consideration and make them as easy as possible to complete.

Storage is an important consideration in efficient design. Be honest about how much storage capacity you will need. There is no point in planning for less storage than necessary — it will be a constant frustration in the home. The tiny house movement has introduced a wide array of clever storage options. Think about spaces in your design that are underutilized; it may be possible to integrate storage into them in unique ways.

Don't strain to be original

You don't have to come up with your own design from scratch. By copying or slightly altering an existing design that has many of the features you want, you can still create a highly personalized living space. Even two houses built from identical plans can look and feel remarkably different and original. Designing a home is not an art competition or exercise in radical originality (unless you want it to be!). Borrowing or even directly copying what has already been done is not cheating — it's a time-honored tradition.

From design to plans

Once you have created or found a design that you feel suits your needs, it will need to be translated into detailed plans. Plans have their own language, a collection of symbols and drawing conventions that allow builders to understand plans from any designer, architect, or engineer.

Every set of home plans contains numerous "sheets" that each convey a specific type of information:

- **Site plan** — Shows the property boundaries and the building location on the property. Services such as sewer, water, and power connections are typically shown, as well as elevation marks to show drainage. Parking, walkways, and other elements may be included.
- **Floor plan** — Shows a fully dimensioned view of the building as if the roof were lifted off. Includes location and dimensions of all permanent elements, including the placement of all walls, doors, windows, stairs, counters, closets, appliances, and fixtures. For multi-level homes, there will be a separate floor plan for each floor.
- **Elevations** — Shows the exterior of the home from all four sides, with no perspective. Includes overall dimensions, roofing and siding materials, door and window locations, and finished soil grades.
- **Sections** — Shows the building as if it has been sawn-through vertically to reveal the details of the roof, walls, floors, and foundation assemblies. Multiple sections may be included to ensure that all assembly types and conditions are shown.
- **Details** — Shows expanded drawings of any floor plan or section elements that require greater detail to understand. Often used to highlight specific structural connections or thermal/air/vapor control layer placement.
- **Mechanical drawings** — Shows the placement of electrical, plumbing, heating, ventilation, and air conditioning (HVAC) systems and routing. It is typical for each system to have

its own drawing page. In some jurisdictions, HVAC drawings are required to include calculation sheets to show that the systems meet code requirements.

Do-it-yourself plans

If you intend to perform your own translation from design to finished plans, it is crucial to obtain a copy of your local building code. The code will outline many parameters required for your plans, including minimum room sizes, ceiling heights, door widths, window sizes, hallway and stair widths, and more. Incorporating code requirements into a set of plans is a major undertaking, and it is the main reason homeowners hire a professional to complete their plans.

If you are new to drafting plan sets, consider learning how to do it. There are many community college courses that introduce basic drafting skills, typically in a particular software program. It will take some time and patience to acquire these skills, but the deeper understanding of the building — especially if you are going to be leading the construction work — can be very rewarding.

Purchasing plans

The internet provides a plethora of ready-made home plans that you can purchase. These can be reasonable options, but it is important to note that there are code variations between jurisdictions; if the plan set you are buying does not specifically claim to meet all the code requirements for your location, it probably does not. Plans that are not drawn to meet local codes will need at least some revision (and sometimes lots of revision) to be brought up to the required standard.

Readers of this book may be considering materials, assemblies, and/or mechanical systems that are not conventional. Purchased plans can likely be adapted to work for your chosen materials, but a knowledgeable practitioner should do this adaptation.

Design professionals

It may make sense to hire a design professional to help you turn your design into a set of plans. See Chapter 8 Creating a Design/ Build Team to examine the various professionals you may want or need to include on your team.

Plans and budgets: The back-and-forth

There is a common conundrum in the home-building process: It is impossible to get accurate cost estimates for your home until you have detailed plans, but once you have detailed plans it can be overwhelming to have to change them to meet financial realities. This budgeting reality needs to be built into your planning schedule — it should not come as a surprise, but anticipated in the overall arc of the design process.

It is not unusual for plans to go through two or three revisions in order to have them meet a budget, though a skilled design team should be able to keep any gaps between anticipated costs and quoted costs to a minimum. Be sure to build this phase into the scope of work for your design team so that each team member knows how much detail to provide before waiting for pricing. Final drawings — especially those that are getting a professional's seal or "stamp" — shouldn't be produced until the budget is approved.

Depending on the extent of the difference between the desired budget and the actual quotes you receive, the revisions to your plans may be minor or extensive. It is best to rework the plans until they are demonstrably within your budget range. Though it may seem like a delay and a hassle to re-draw and re-quote the building, this is better than proceeding on to permits and construction with a mismatch between the plans and the budget. While it may be possible to reduce costs on the fly by choosing less expensive materials or finding cheaper labor quotes, this tactic rarely works out very well.

The thrill of a "paper house"

It is exciting to see your house go from an idea in your head to a detailed set of drawings. The project will take on an air of reality that may have been lacking during the design stages, but with a completed set of plans, you will be ready to apply for a building permit and start construction!

Chapter 8

Creating a Design/Build Team

THE ACT OF BUILDING A HOME that is ecologically sound — combining high efficiency with low-impact materials — is not likely to be a solo venture. Although no element of creating such a home is beyond the capability of a determined homeowner to learn and implement, it is much easier to accomplish with the assistance of knowledgeable and experienced professionals. Within the design profession, people with specific skill sets are realizing the advantages of partnering with others who have complementary skills and approaches. The best buildings being made today are coming from integrated design teams, and assembling your own team is the best way to ensure that you will meet all your goals.

Who might be involved?

Design teams can vary based on the needs of the homeowner and the skill sets of each team member. Below are listed the key types of knowledge that you want team members to have, along with suggestions regarding the people who may have that knowledge:

1) Architectural

In most jurisdictions, there are a number of possible pathways for obtaining architectural services for residential construction.

Skill set includes:

- **Design services** — Help to create an overall design.
- **Architectural rendering** — Creating complete plan sets.
- **Building code compliance** — Ensuring plan sets meet all code requirements.

Can be provided by:

- **Architect** — Licensed by a governmental body to practice in their state or province and take legal responsibility for plans.
- **Architectural technologist or draftsperson** — Trained to draw plans and in some jurisdictions and conditions can take legal responsibility for plans.
- **Engineer** — Licensed by a governmental body to practice in their state or province, and take legal responsibility for plans.
- **Builder** — Design/build firms can provide architectural services and, within limits take legal responsibility for plans.
- **Homeowner** — In many jurisdictions, within limits, homeowners are able to submit and take legal responsibility for their own plans.

2) Structural

Residential construction does not necessarily require structural design by a design professional. If the building is designed within the prescriptions of the building code, it is assumed to have adequate structural integrity without requiring a professional to verify. However, if the plans call for materials or assemblies that are considered "alternative," additional structural design by a design professional may be required.

Skill set may include:

- **Structural design** — Ensuring plans meet all code requirements for structural integrity.
- **Material verification** — Ensuring materials meet structural integrity requirements.
- **Inspection** — Ensuring that on-site construction meets all design criteria.

Can be provided by:

- **Structural engineer** — Licensed by a governmental body to practice in their state or province, and take legal responsibility for the structural integrity of all or part of the plans.
- **Architect** — Licensed architect can take legal responsibility for residential structural design, or may offer engineering services within the firm.
- **Engineering technologist** — Trained in structural design; in some jurisdictions can take legal responsibility for structural design of all or part of the plans.

3) Energy modeling

In many jurisdictions, energy modeling is required to meet basic code requirements, and it is certainly necessary to meet rating system (for example: LEED, Passive House) requirements. For any homeowner attempting to exceed code minimum standards, it is certainly recommended. There are many different software programs for energy modeling, and matching software with code and/or rating system requirements is important.

Skill set may include:

- **Basic energy modeling** — Some energy modeling programs offer a limited number of variables and provide the basic information required to meet code requirements.
- **Advanced energy modeling** — Some energy modeling programs offer a wide range of variables (including passive solar, thermal mass, thermal bridging, comfort analysis) and can provide information required to meet certain rating system requirements.
- **Hygrothermal modeling** — Some modeling programs offer the ability to model both heat and moisture flows and can provide helpful information for controlling potential moisture issues in high-performance buildings.

Can be provided by:

- **HVAC engineer** — Licensed engineer specializing in the design of HVAC systems will be able to offer energy modeling services.
- **Architect** — Licensed architect may be able to offer energy modeling services within firm.
- **Building scientist** — Trained and/or licensed building scientist will be familiar with energy modeling programs.
- **Rating system provider** — Many rating systems offer certifications to providers who will manage the delivery of the rating program. This provider may be able to offer energy modeling services to meet the needs of the program.
- **Builder** — Design/build firms may be able to provide energy modeling services.
- **Homeowner** — Homeowners may be able to learn appropriate software program and provide his/her own energy modeling.

4) HVAC design

Most jurisdictions require the design of heating, ventilation, and cooling systems to be done by a qualified person, usually to a prescribed standard in the code.

Skill set may include:

- **Energy modeling** — HVAC design and energy modeling are often undertaken in conjunction with one another.
- **System design** — Providing complete plans for systems to meet code requirements and overall efficiency targets.
- **Equipment specification** — Providing a list of equipment capable of meeting code and building needs.

Can be provided by:

- **HVAC engineer** — Licensed engineer specializing in the design of HVAC systems.
- **Architect** — Licensed architect may be able to offer HVAC design services within firm.

- **Building scientist** — Trained and/or licensed building scientist may be may be able to offer HVAC design services.
- **HVAC installer** — Companies specialized in HVAC installation may be able to offer HVAC design services.
- **HVAC equipment providers** — Manufacturers or distributors of HVAC equipment can often provide design services for their own products.
- **Builder** — Design/build firms may be able to provide HVAC design services based on code provisions.
- **Homeowner** — Homeowners may be able to provide his/her own HVAC design based on code provisions.

5) Building science design

Residential design typically does not require plans to be reviewed by a building scientist. However, homeowners attempting designs with exceptional energy efficiency and/or innovative materials will benefit from having an analysis done by somebody trained in building science.

Skill set may include:

- Energy and moisture modeling
- Airtightness planning
- Durability planning

Can be provided by:

- **Building scientist** — Trained and/or licensed building scientist may be able to offer HVAC design services.
- **Architect** — Licensed architect may be able to offer building science services within firm.

6) Project manager

The project manager leads the design and build team, coordinating meetings and the exchange of information and drawings between team members. The project manager may have one or more active roles on the team as well.

Skill set may include:

- **Strong commitment to project goals** — Responsible for ensuring goals are met by team.
- **Good interpersonal skills** — Efficient at running meetings, coordinating efforts, resolving disputes.
- **Budgeting** — Leads or assists in ensuring design decisions match budget expectations.
- **Design/build skills** — Understands phases of design and construction and overall project arc.

Can be provided by:

- **Any team member** — Homeowners often choose to take this role, but it can be provided by any team member (architects and general contractors often take this role) or by someone with no other role in the design/build team.

7) General contractor

The general contractor is responsible for coordinating all on-site construction activities, and is often a participant in design team meetings.

Skill set may include:

- **Strong commitment to project goals** — Responsible for ensuring goals are met by build team.
- **Good interpersonal skills** — Efficient at coordinating suppliers and trades, resolving disputes.
- **Budgeting** — Directly handles the purchasing of materials and payment of tradespeople.
- **Design/build skills** — Understands phases of construction and overall project arc.

Can be provided by:

- **Licensed general contractor** — Most jurisdictions have a legal definition and requirements for a licensed general contractor.
- **Design/build firm** — Design/build firms may be able to provide general contracting services.
- **Homeowner** — In most jurisdictions, homeowners are able to submit plans and take legal

responsibility for construction of their own home.

8) Other participants

There are numerous other participants who may play an important role in your project. It can be valuable to have these people participate in at least one team meeting to gather their input for the design:

- **Renewable energy designer/installer** — Projects that will incorporate renewable energy systems will benefit from having an experienced designer (often also the installer) at the table to ensure that system sizing, integration, and budget meet project goals.
- **Electrician and plumber** — Though not typical, it can be very helpful to have the project electrician and plumber attend a design meeting to ensure that they understand the project goals and provide input regarding design decisions that affect their role.
- **Building biologist** — If your goals involve an emphasis on indoor environment quality, involve a building biologist or a consultant with appropriate knowledge about air and water quality, electrical fields, and healthy material choices.
- **Rating certifier/provider** — Each rating system (see Chapter 4) will have requirements for meeting their certification thresholds. In some cases, the rating certification must be handled by a third party who is not directly involved as a member of the design team. In other cases, the rater can play an active role with the design team. In almost all cases, certification requires proof of a team-based design approach.
- **Specialist consultants and/or installers** — A particular material or system may require specialist knowledge that doesn't warrant a role on the design team, but can provide a key contribution to the team.

Assembling the team

Team-based design and construction is a relatively new approach, especially in residential building. There are very few firms that offer all of the services described above entirely in-house. In fact, it is very unlikely that many of the professionals who offer these services have ever been asked to sit together in a collaborative effort. This puts the onus for assembling the team on your shoulders. You won't necessarily need all of the people listed above to be on your team, as some members will be able to fulfill multiple roles. And some members may only need to be peripherally involved, providing key input at a particular juncture in the process, but not taking part in the overall planning.

First and foremost, you should feel very comfortable with all the team members. Personality is at least as important as a consultant's resume! You will be working closely with these people, sharing your hopes and dreams for your home and involving them in major personal and financial decisions. It is not impolite to ask for references from past clients and to follow up with the references. You will be making a significant time and financial commitment to your team members, and you should feel confident that you can get along, work together creatively and productively, and resolve issues constructively.

Why do this?

Residential construction is a rapidly changing field. Performance expectations are increasing, material options are expanding, energy sources are in a period of diversification, health concerns are growing. It is increasingly unlikely that any one person has enough experience and training in all of these areas to meet high goals. The addition of different perspectives can help to identify potential oversights or problems in the plans and smooth them out at the design stage — before they can cause issues or delays on the construction site.

All of the "new" criteria addressed in Chapter 5 add elements to the design process that have not typically been addressed by the industry or in professional training. The best way to meet comprehensive criteria is to involve people who have experience in meeting each of those criteria.

A team effort can also help to keep the design process on track, with team members responsible for providing their input and deliverables according to the team's schedule.

There is no sure-fire template for coordinating an integrated design team. The need for the whole team to meet at once is quite limited, and may only need to happen once, near the beginning of the project. The design will be iterative, with each person's contributions paving the way for another person on the team to undertake some or all of their scope of work. Many of the discussions after the first meeting can happen directly between team members; the coordinator is the one charged with being certain that the necessary information is making it to appropriate team members.

It is likely to cost some extra money in the early stages to engage the entire team in the design process, but experience among integrated design teams indicates that this approach typically saves a great deal of money; the result is better design solutions, improved building efficiency, and minimized oversights that end up costing money on the job site.

Using the Criteria Matrix

The Criteria Matrix presented in Chapter 5 was developed to play a central role in the work of the design/build team. A suggested order of operations would be:

- Present draft matrix to lead consultant — the team member who will have the widest range of responsibilities identified in the matrix (general contractor, project coordinator, architect, or homeowner).
- Collaborate to finalize the matrix and set identifiable and definitive targets for each criteria.
- Coordinate a list of additional team members required to meet all the criteria.
- Grow the design team by recruiting members who clearly understand the criteria targets being presented and who are enthusiastic about helping to meet them.
- Create a plan for gathering the input of all team members, taking into account areas in which one member may rely on another in order to move forward.
- Use the matrix as a checklist as the design develops, ensuring that the results for each element of the building meet the intended targets.
- If a criteria target cannot be met, decide on a way to resolve the issue. If it is a singular issue, it may just be an anomaly that, upon examination, is allowed to stand; if one criterion continually presents issues and is difficult or impossible to meet, the target may need to be adjusted.
- Use the matrix to complete a final review of the plans before they are submitted for permit.
- Present the design matrix to all those who will be working on the building site and be sure that they understand which criteria they have a role in meeting and what targets their work is intended to reach.
- Share the design matrix with key suppliers so they understand your intentions and direct you to the most appropriate materials.
- When the project is complete, do a review to verify that all your goals were met. Discuss with team members any goals that were not met, so everybody can learn where the shortfalls occurred and how to avoid them in future projects.

Clear goals are the best means to ensuring that the whole team is working together coherently, and the Criteria Matrix is a straightforward way to present your goals and keep them as the central focus for the project. Of course, the matrix can be modified to suit the needs of your project — some criteria can be eliminated, others added. The principle of a clearly defined set of goals is the intention; if this matrix can help you and your team to achieve this, that's great. If not, develop your own format and use that.

In the end, the result should be a skilled team applying their efforts to a well-defined end. If you can achieve that, your project will be a success.

Chapter 9

The Regulatory System

ANY HOMEOWNER OR BUILDER faces a potentially intimidating regulatory process when undertaking a building project. There can be a lot of red tape to cut through on the way to erecting a building. Though most homeowners are preoccupied with simply obtaining a building permit, the permitting process can also involve some — or all — of the following:

- **Zoning regulation** — Land use acts define what types of development can be done on a property, and residential construction is not allowed in all areas. Before purchasing a property, be sure to understand the zoning regulations that apply. It is possible to get zoning regulations changed, but the process can be slow and expensive, and there is no guarantee that changes can be made.
- **Development permits/fees** — Many jurisdictions charge fees associated with developing a property. In some urban areas, this can amount to tens of thousands of dollars. This permit ensures that the municipality has the capacity to service the property (provide road, sewer, water, power). If the zoning for the property is appropriate, these development permits are typically not difficult to obtain, but may be expensive and should be figured into the project budget.
- **Conservation authority approval** — Some jurisdictions give control of development in watersheds and other ecologically sensitive areas to some form of regulatory body. If your property falls within their boundaries, a permit may need to be obtained and a fee paid.
- **Heritage act permit** — Extensive renovations to historical buildings or new developments on or adjacent to historical lands may require some form of heritage act permit be obtained and a fee paid.
- **Septic permit** — For properties unserviced by municipal sewer systems, a permit must be obtained for a private septic system. This permit is often required to be obtained before a building permit application is made. A licensed designer may be required, and a fee paid.
- **Entrance permit** — Access to roads must be approved for new property development (just because a driveway exists doesn't mean it's an *approved* driveway). A permit may need to be obtained and a fee paid. The permit will depend on legal conditions being met for sightlines, distance from intersections, road speeds, shoulder width, and other considerations. There is no guarantee that an entrance permit can be obtained.
- **Building permit** — Most jurisdictions require a building permit and a series of inspections to fulfill the permit. See below for more detail.
- **Electrical and/or plumbing permit** — In some jurisdictions, electrical and/or plumbing permits and inspections may be handled by a different agency than the building permit. Separate fees may also apply.
- **Occupancy permit** — When construction is completed, the building department may make a final inspection (and may request to see any or all of the above permits) before issuing an occupancy permit. This may be used to initiate property taxes and trigger insurance and mortgage conditions.

As this list implies, there can be a long pathway between you and the construction of your home. Take the time to investigate the regulatory regime in your region, and be sure to include the time and cost estimates for each permit required in your overall project planning. Some jurisdictions offer a document that guides would-be home builders through the local permitting process, and these can be invaluable.

Building permits

Despite the long list of potential regulatory hurdles to be crossed, we will focus here on the building permit because of its direct effect on the selection of materials and systems according to your Criteria Matrix.

Don't dread the code

There is commonly a sense of dread that hopeful homebuilders have in regard to building codes and obtaining permits. Everybody has heard a horror story or two about permit denials and protracted struggles to get a building permit. While such cases do sometimes occur, they are not typical, and even builders using alternative types of materials and systems are able to obtain their permits with relative ease. This chapter will help you to prepare adequately for your own permit application.

Despite plenty of anecdotal evidence to the contrary, there is no legal justification in any North American building code for denying a permit because of the use of alternative methods and materials. At the same time, it is highly unlikely that any building department that will respond to a general inquiry about whether or not they will permit the use of a particular alternative with a general answer of "yes." Permits are not granted or denied on the basis of a single material choice, but for meeting a complete set of requirements that demonstrate the viability and safety of the entire structure.

Model codes and local codes

Both the International Residential Code (IRC) in the United States and the National Building Code (NBC) in Canada are *model codes.* They are written to cover a broad range of conventional practices and solutions applicable across each country. These model codes are then adopted by state, provincial, county, and/or municipal authorities, and these levels of government can adopt and modify the model code to suit local needs. The first step in working with the building code is being sure which code you need to work from (and which is the most up-to-date version of that code in your jurisdiction). If you are intending on playing any key role in your own design team, it is well worth your while to obtain the pertinent copy of the code and become familiar with how the code is structured and how to find the information you need to make informed decisions.

Approved or accepted solutions

Model codes are an accumulation of "accepted solutions" that prescribe how buildings can be designed and constructed in order to meet minimum standards of occupant and community safety. A building permit will be issued if a set of plans meets all of the prescriptions in the relevant sections of the code.

If you selected "1-Fully Code Compliant" in your Criteria Matrix, then you will want to ensure that that all of your design, material, and systems decisions are in conformance with the prescriptive language of the code. There are multiple options and pathways embedded within the code's prescriptions, and you and your design team can work through these prescriptive options and find the ones that best match the rest of your criteria choices. It may be possible to meet high goals in each category of the Criteria Matrix using fully code compliant prescriptions.

Alternative solutions

Given that sustainable building criteria are a relatively new consideration in the building industry, a lot of options that meet high sustainability goals may not be directly recognized in the prescriptive language of the building code.

Every building code has a mechanism for consideration of non-conforming materials and approaches. If you have concluded that some elements in your design are not supported via code prescriptions, it is incumbent on you to understand the exact procedures used in your code jurisdiction to handle alternative compliance. While the paperwork requirements may vary (and additional permit fees may apply), all such alternative compliance pathways operate on the assumption that the applicant will provide proof that the alternative proposal meets or exceeds the provisions of the prescriptive code requirements. Any performance parameter (structural capacity, fire resistance, thermal performance, etc.) that exists for a building component in the prescriptive section of the code must be demonstrably met or exceeded by the proposed alternative. Each of these performance parameters must be fully supported and documented.

Table 9.1 includes the alternative compliance language from both model codes, which share much in common. From these excerpts, you can determine the type of proof of equivalency required to meet alternative compliance expectations.

Table 9.1: Building code chart

USA — International Residential Code (IRC-2015 edition)	Canada — National Building Code (NBC-2012 edition)
R104.11 Alternative materials, design and methods of construction and equipment. The provisions of this code are not intended to prevent the installation of any material or to prohibit any design or method of construction not specifically prescribed by this code, provided that any such alternative has been *approved*. An alternative material, design or method of construction shall be *approved* where the *building official* finds that the proposed design is satisfactory and complies with the intent of the provisions of this code, and that the material, method or work offered is, for the purpose intended, not less than the equivalent of that prescribed in this code. Compliance with the specific performance-based provisions of the International Codes shall be an alternative to the specific requirements of this code. Where the alternative material, design or method of construction is not *approved*, the *building official* shall respond in writing, stating the reasons why the alternative was not *approved*. **R104.11.1 Tests.** Where there is insufficient evidence of compliance with the provisions of this code, or evidence that a material or method does not conform to the requirements of this code, or in order to substantiate claims for alternative materials or methods, the *building official* shall have the authority to require tests as evidence of compliance to be made at no expense to the *jurisdiction*. Test methods shall be as specified in this code or by other recognized test standards. In the absence of recognized and accepted test methods, the *building official* shall approve the testing procedures. Tests shall be performed by an *approved* agency. Reports of such tests shall be retained by the *building official* for the period required for retention of public records.	Documentation of Alternative Solutions **2.1.1.1. Documentation** (1) The person proposing the use of an alternative solution shall provide documentation to the chief building official or registered code agency that, (a) identifies applicable objectives, functional statements and acceptable solutions, and (b) establishes on the basis of past performance, tests described in Article 2.1.1.2. or other evaluation that the proposed alternative solution will achieve the level of performance required under Article 1.2.1.1. of Division A. (2) The documentation described in Sentence (1) shall include information about relevant assumptions, limiting or restricting factors, testing procedures, studies or building performance parameters, including any commissioning, operational and maintenance requirements. 2.1.1.2. Tests (1) Where no published test method to establish the suitability of an alternative solution proposed under Article 2.1.1.1. exists, then the tests used for the purposes of that Article shall be designed to simulate or exceed anticipated service conditions or shall be designed to compare the performance of the material or system with a similar material or system that is known to be acceptable. (2) The results of tests or evaluations based on test standards other than as described in this Code may be used for the purposes of Sentence (1) if the alternate test standards provide comparable results.

In both codes, several options exist for demonstrating that an alternative solution meets or exceeds code requirements:

- **Testing data** — Both codes specify a preference for tests that are done to a code-recognized standard, such as ASTM, ANSI, or CSA. If the tests were not performed to the standard used by the code, you will be required to show how the testing varies from these standards and how the results may be interpreted to show equivalency. For example, if you are using tests performed to European standards, you will need to show how they meet the intent of North American tests and standards and account for any differences in testing methodology and results.
- **Referenced standards** — Some materials and assemblies are not directly recognized as accepted solutions within the code, but an independent standard may exist that is referenced by the code. In these cases, a product manufacturer or trade association will have hired a standards organization (such as ASTM, CSA, CCMC [Canadian Construction Materials Centre], or others) to create a standard that can be followed by a designer and/or builder.
- **Past performance** — An applicant can cite prior examples of the same (or similar) approach used successfully in the jurisdiction. Be sure to have adequate documentation of past performance to ensure that the approach was similar to what you are proposing, and to be sure that it was indeed a successful approach. Past performance or case studies from other jurisdictions or other countries may not be viewed as conclusive evidence, especially if the climates are different. The quality of documentation will be examined carefully, and if it doesn't stand up, it may not be recognized as proof of equivalency.
- **Professional seal** — Though not directly referenced in the code excerpts above, a licensed architect and/or engineer can often provide code equivalency assurance to the building department by applying their seal to the drawings and so ensure that to the best of their professional ability the alternative approach meets the intent of the code.

It may be that all of these approaches are employed on an alternative compliance application. It is entirely up to you as the applicant to provide documentation and any supporting interpretation to the building department. For better or for worse, building departments are reactive, not proactive. They are under no obligation to assist you with your documentation or ensure that it is complete. They are only obliged to respond to what has been provided in the application.

Code consultants are professionals that can be hired to assist an applicant with understanding the code and all the parameters that need to be addressed in order to put forward a complete application. The code consultant can just consult, or he or she can handle the whole submission.

Rejections and appeals

It is important to know that a permit cannot be denied for any reasons other than code infractions or incomplete submissions. Every building code prescribes the manner in which a denial is presented to the applicant. In most code jurisdictions, the procedure for a permit refusal involves a written response explaining the code infractions that caused the permit to be denied. This is intended to give the applicant a full understanding of where the application was found to be lacking and provide a blueprint for resolving the issues in a resubmission. If all of the code issues are fully addressed in a subsequent submission, then a permit should be issued. In many cases, there can be several rounds of rejection and resubmission. While this may be frustrating and time consuming, getting a building permit can be compared to taking a test where you must score 100%. Building departments cannot let any infractions they detect slip through without being addressed, so it is best to consider the application to be a multi-step affair. Forming and maintaining a good working relationship with the plan reviewer is very helpful. At best, the plan reviewer will be acting as an advocate and will be assisting you with understanding where the plans fall short of meeting the code and making suggestions regarding how the deficiencies can be corrected. At worst, they are obliged to make your mistakes known to you, and you will have to figure out how to correct them.

Building codes cover a wide range of issues and topics, and there are many areas in which a set of building plans may not conform to the code that have nothing at all to do with alternative compliance issues. In the author's experience as a consultant, the majority of permit denials for projects proposing "alternative" materials have to do with issues that are unrelated to these specific material choices; denials are much more likely to be related to zoning issues (lot lines and setbacks, overall height, grading, parking allocation), space allocation (minimum room sizes, means of egress, staircases and railings, window size and placement), and services (well/water, sewer/septic, HVAC) — none of which have anything to do with alternative materials or assemblies. These are common problems experienced by conventional and alternative proposals alike. Addressing them requires an understanding of the codes, but does not directly influence the use of alternative approaches.

Should there be a disagreement about code compliance, every code jurisdiction has an established route for appeals. Often, this involves taking the dispute to the Chief Building Official. Should this fail to resolve the issue, there will be a higher regional, state, or provincial authority that will hear appeals, and the pathway to accessing the appeal should be provided to you. Many appeal processes are quasi-judicial and involve a hearing where both the applicant and the building department put forth their arguments and a panel renders a decision.

The appeal process can add time and cost to a project, but it also tends to resolve matters in favor of a reasonable application. Many building departments will take a matter to appeal not to prevent the building project from going forward, but to receive a directive ruling from a higher authority. This can help to deflect potential liability issues for the municipality and clarify the local building department's interpretation of the code.

Preparation and patience are invaluable

Any application to a building department involving an alternative compliance element should be made well in advance of needing the permit to allow the process to go through a few rounds of back-and-forth. Expecting or, worse, demanding a fast turnaround for an alternative compliance application is an invitation to frustration and delays.

Any applicant willing to put the time and effort into making a viable and complete initial submission and diligent enough to follow through with any requests for changes or more information can expect to be rewarded with a building permit.

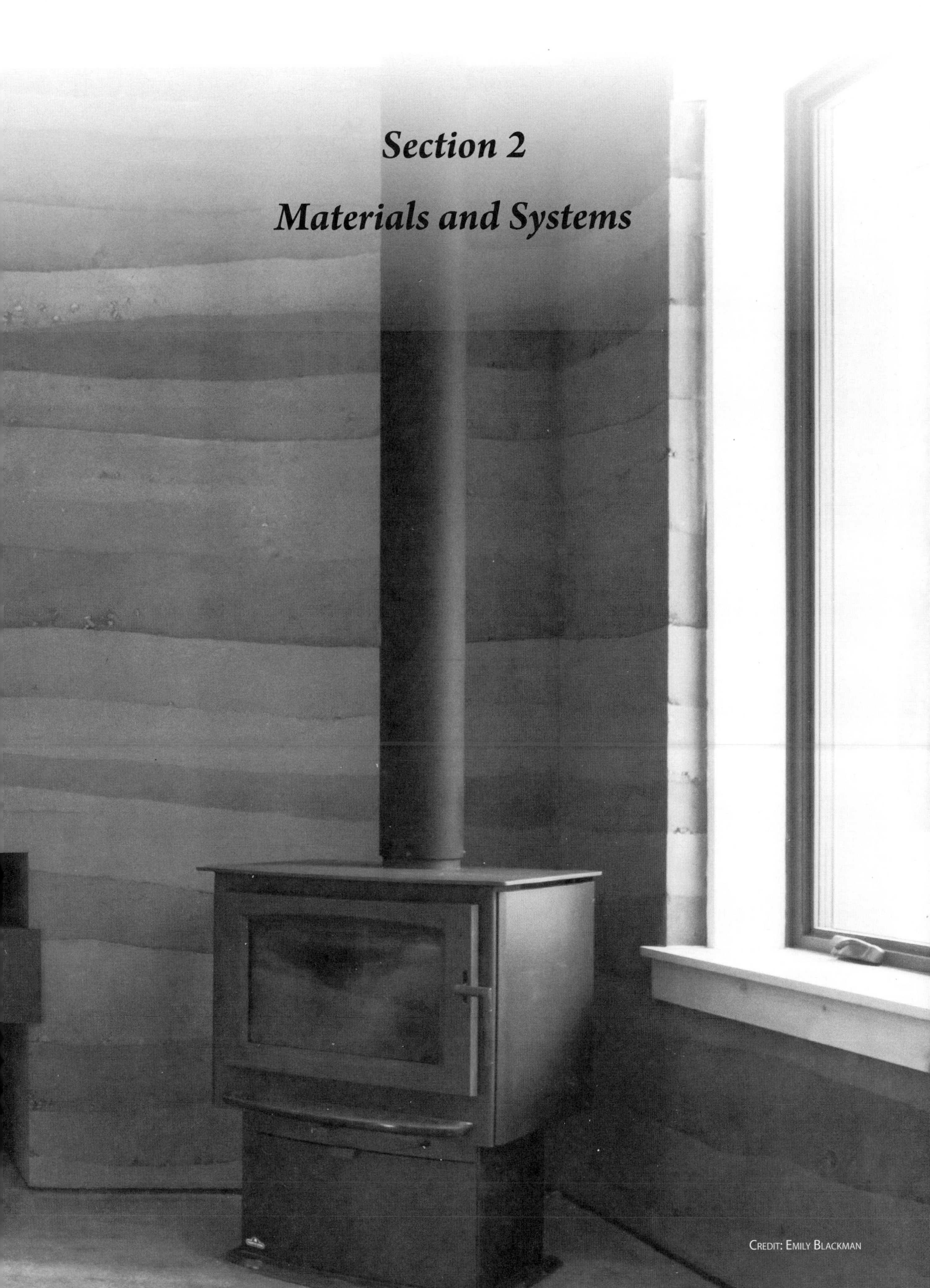

Section 2

Materials and Systems

Credit: Emily Blackman

Chapter 10

Structural Materials

THIS BOOK HAS ENCOURAGED you to generate a Criteria Matrix for your project and to establish an understanding of the building science and building code implications of different design approaches. Now it is time to consider the wide array of options for the structural and mechanical components of your building, and to begin the process of selecting those most appropriate for your project.

This section of the book categorizes materials and systems and gives a brief synopsis of many of the most sustainable options. *Criteria Considerations* are included to help you identify the options that most closely match your overall goals. Key selection criteria include:

- Ecosystem impacts
- Embodied carbon
- Energy efficiency
- Indoor environment quality (where applicable)
- Waste
- Building code compliance
- Resilience (where applicable)

The Criteria Considerations can give you a strong indication of which choices are worth exploring in more detail, and which may not be. Specificity is crucial in making final selection. Not every criterion included in the Criteria Matrix is considered for each material and system option, only those that are most relevant.

While material and system selections will certainly have major implications for all of these criteria, the overall goals are not affected on a material-by-material basis. It is possible to build a low-maintenance, low-cost, and/or low-labor house using any of the materials and systems presented here. Meeting your goals in these areas is a matter of creating an entire assembly that is suitable — the role of any single material or system in that effort will not determine the overall range in which the whole project will fall.

Not every material and system is covered here, and none are covered in detail. Here, you will find enough information to help you decide which options might be right for your project, and you'll find a list of resources at the back of the book that will help you learn about each one in more detail.

Structural Materials: Foundations

A building's foundation is extremely important to its longevity and performance. As such, it is often the one element where homeowners and builders will tend to choose the tried-and-true techniques and avoid "experimentation."

This is unfortunate, because the conventional methods and materials typically involve the highest environmental impacts and often the lowest energy efficiency. Most North American homes use vast amounts of concrete in their foundations, and concrete is a perfect example of the kind of energy-intensive building material that has led us to our current environmental state. The production of the Portland cement that is the "glue" in concrete requires using large quantities of fuel to heat limestone to very high temperatures to change its chemical composition. In the process, the carbon dioxide trapped in the stone is released into the atmosphere (along with additional CO_2 released by the fuel used to heat the rock). Cement manufacture is one of the world's leading sources of greenhouse gas emissions. Concrete foundations are commonly paired with the use of petrochemical foam insulation and waterproofing materials, which furthers the ecological and health damages associated with conventional foundation options.

In considering more sustainable foundation systems, a builder is forced to consider a number of challenges to typical expectations. In many parts of North America, foundations have been twinned with conditioned, subgrade living space: the basement. In those markets, having a basement is so normal that it can be hard to convince a homeowner to imagine a house without one. Creating a basement foundation almost always requires the use of materials that are not sustainable, and if you have high goals for embodied carbon and ecosystem impacts, it will be difficult to meet them with a basement. One benefit to moving away from conditioned basement foundations can be reduced cost. The savings that can be realized by using a more sustainable, grade-based foundation are substantial.

As you will see in this section, there are many ways to create stable, long-lasting foundations that have reasonable environmental impacts. Many of them, however, do not make basement foundations, and those that do come with significant labor requirements. The fact of the matter is that building large, conditioned basements has been a privilege of having cheap energy at our disposal and no concern for climate change. We are nearing the end of commanding that privilege.

It is important to remember that the various foundation systems included here will likely require water control and thermal control layers that are not inherently part of the foundation, and that the materials used for these layers must also be considered against your Criteria Matrix in order to rate the foundation system as a whole.

There is no doubt that the most skepticism and wariness about sustainable technologies will happen here, at the foundation. As with any change, the underlying assumption — the "foundation" — is the hardest to change. Yet this is the place that most needs changing.

Concrete

How the system works

Concrete combines two main ingredients: aggregates (sand, gravel) and a hydraulic cement binder, most commonly Portland cement. The ingredients are mixed with water, which begins a chemical reaction with the cement binder that hardens and glues the aggregate together. Formulations for concrete vary based on many factors, including requirements for strength, plasticity, setting time, and environmental conditions.

Concrete in residential construction is used in three different forms:

- **Poured concrete** — Temporary formwork is constructed and concrete from a batching plant is brought to the site pre-mixed and poured into the forms. In some cases, builders will mix the concrete ingredients on site. Once the concrete is set, the forms are removed.

Applications	Properties
• Piers • Frost walls (incl. basement walls) • Perimeter beams • Slabs • *Can also be used as above-grade walls*	• R-value: *0.1–0.5 per inch* • Compressive strength: *20–40 MPa (3000–6000 psi)* • Density: *960–2400 kg/m³ (60–150 lb/ft³)*

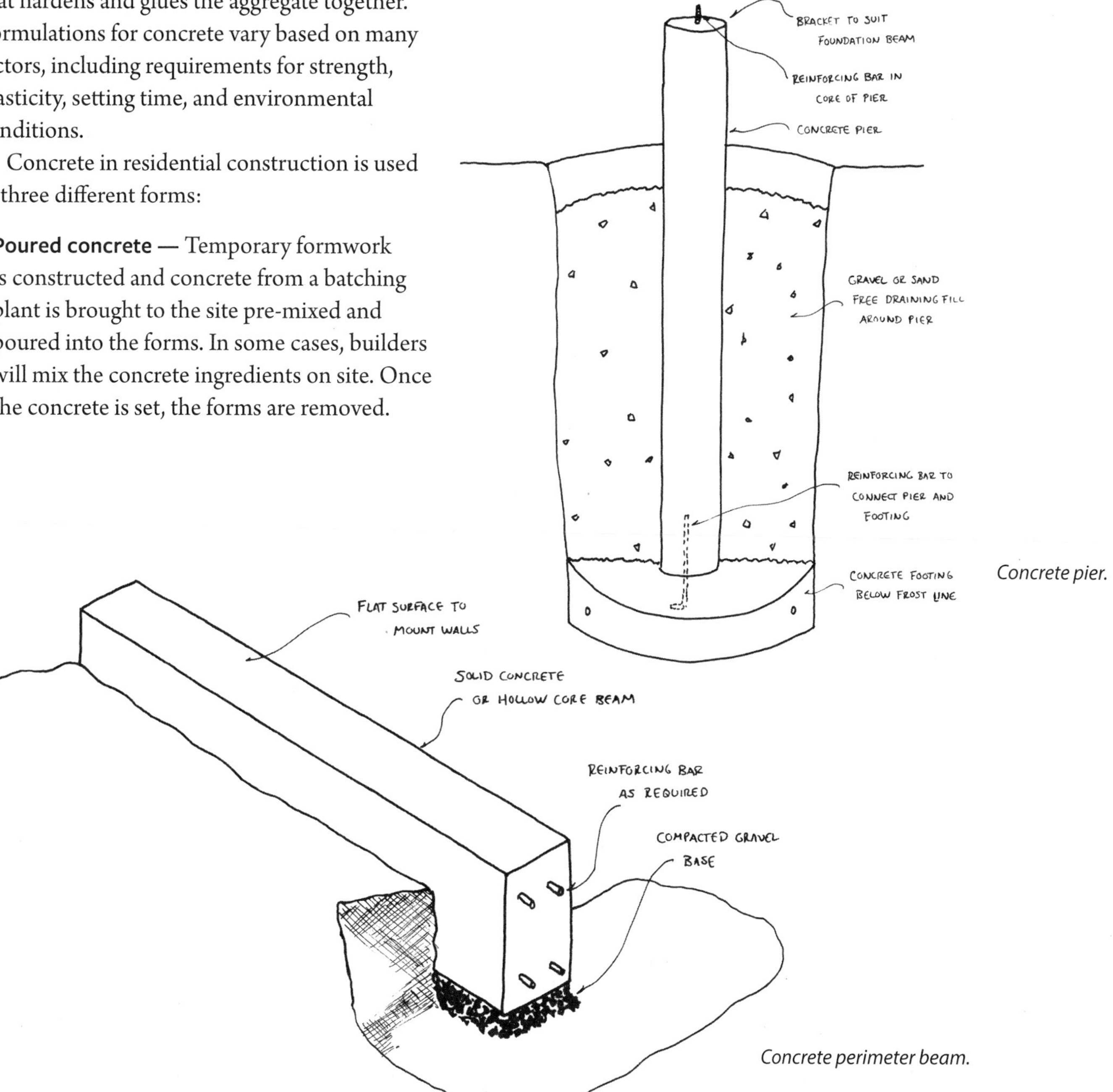

Concrete pier.

Concrete perimeter beam.

- **Pre-cast concrete blocks (concrete masonry units or CMUs)** — Formed and cured off site, CMUs are delivered to a site and mortared together. CMUs feature hollow cores that can be filled with concrete to provide structural support.
- **Pre-cast concrete panels** — Modular panels are custom formed and cured off site and installed by boom truck or crane.

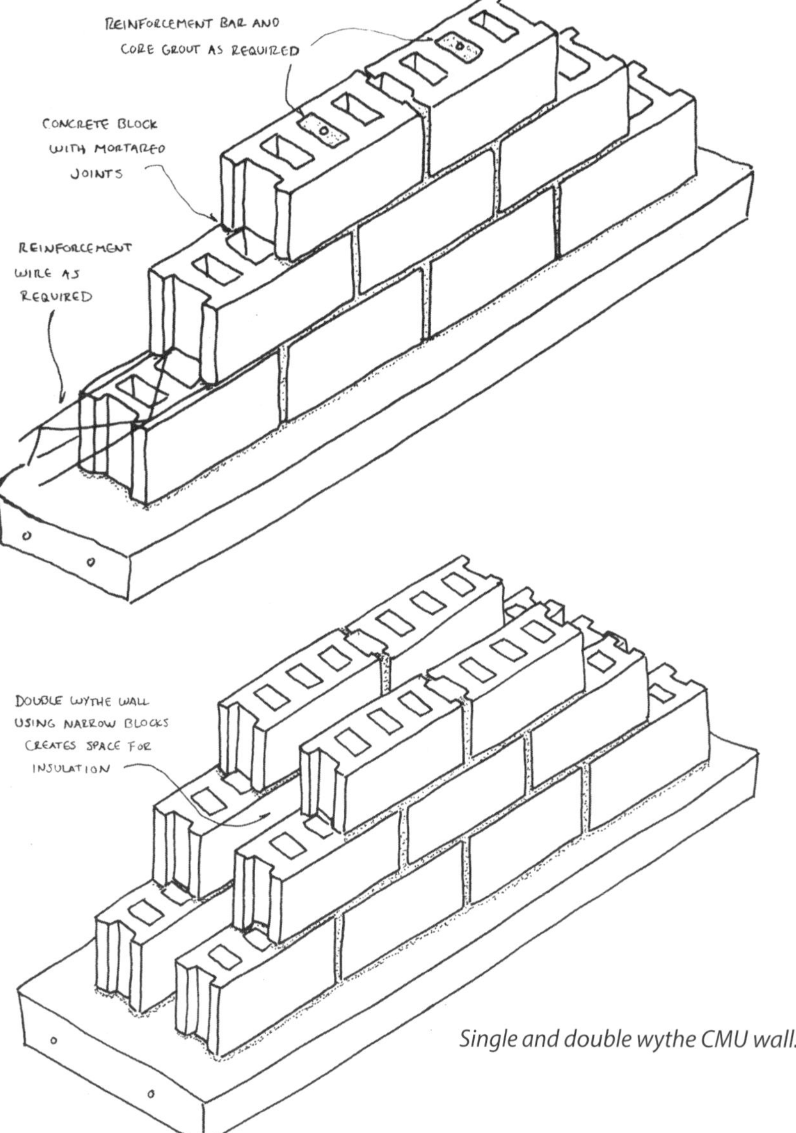

Single and double wythe CMU wall.

Criteria Considerations

 ***Ecosystem impacts:*** Check regional practices for aggregate quarries, as impacts can vary widely. Consider the use of recycled aggregate or urbanite to mitigate impacts.

 ***Embodied carbon:*** Cement production is a major contributor to climate change and concrete will contribute significantly to the carbon footprint of your building. Developments in low-carbon, carbon-sequestering, and microbial concrete may change this, but most of these products are currently prototypes and not widely available.

 ***Energy efficiency:*** Insulation will need to be added to most concrete foundations, and the full criteria implications of the insulation strategy must be carefully considered.

 ***Indoor environment quality:*** Inert. Insulation choices may have IEQ consequences, as will finishes applied to exposed concrete. Waterproofing materials on the exterior may contaminate soil/groundwater.

 ***Building code compliance:*** Code compliant.

 ***Material costs:*** Full-system costing should include formwork construction/removal, reinforcement materials, and insulation.

 ***Labor:*** Contractors and suppliers widely available. Can be formed, mixed, and poured by owner-builder.

Lightweight concrete

How the system works

There are many versions of lightweight concrete, and it goes by many names, including foam(ed) concrete, foamcrete, cellular lightweight concrete, and aerated concrete. These materials can provide the durability, nontoxicity, and strength provided by regular concrete, while also providing some degree of thermal insulation. Typically, higher-strength mixes have less thermal insulation, and extremely light and insulative mixes have much lower strength. In an ideal scenario, the thickness of lightweight concrete required to provide the structural needs of the foundation would also provide the thermal insulation requirements.

Materials in this category fall broadly into three types:

- **Autoclaved aerated concrete** — AAC typically comes as pre-cast blocks or panels. Quartz sand is the aggregate, and Portland cement is mixed with aluminum powder, causing a chemical reaction that creates bubbles that aerate the mix, creating up to 80 percent void space. The mixes can be designed to have specific densities for particular uses. The formed, wet mix is steam-pressure hardened (autoclaved) for up to 12 hours, giving it its structural properties.
- **Foamed concrete** — Foamed concrete can be pre-cast, or installed into site-built formwork. Hydraulic cement and aggregate are mixed into a slurry, and then mixed with an aerating foaming agent that can create bubbles that will remain stable during the pouring, placing, and curing process. The foaming agent can be protein based or synthetic, and resembles shaving cream.
- **Lightweight aggregate** — This type of mixture is most commonly installed into site-built formwork. Hydraulic cement can be used to bind together lightweight, mineral-based aggregates, including naturally occurring pumice, scoria, and diatomite or manufactured aggregates like expanded clay, glass, or perlite.

Applications	Properties
• Frost walls (including basement walls) • Perimeter beams • *Can also be used as above-grade walls*	• R-value: *0.5–2.5 per inch* • Compressive strength: *1–10 MPa (145–1450 psi)* • Density: *150–1600 kg/m³ (10–100 lb/ft³)*

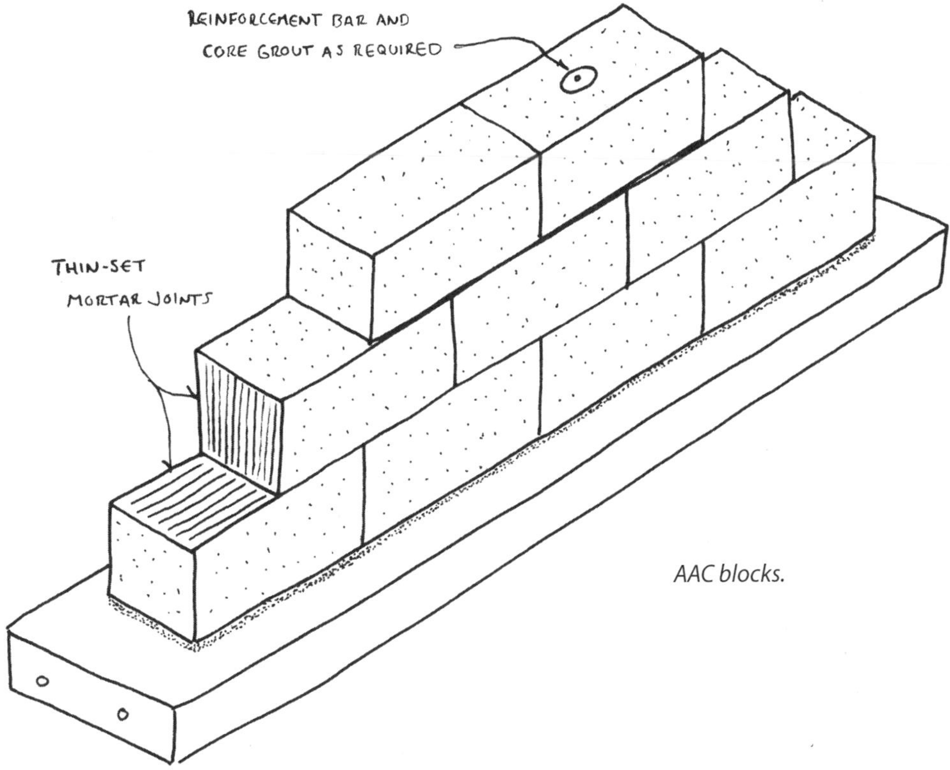

AAC blocks.

Matching a lightweight concrete option with your goals will require research into the specific product type, as there are many variables within this material category.

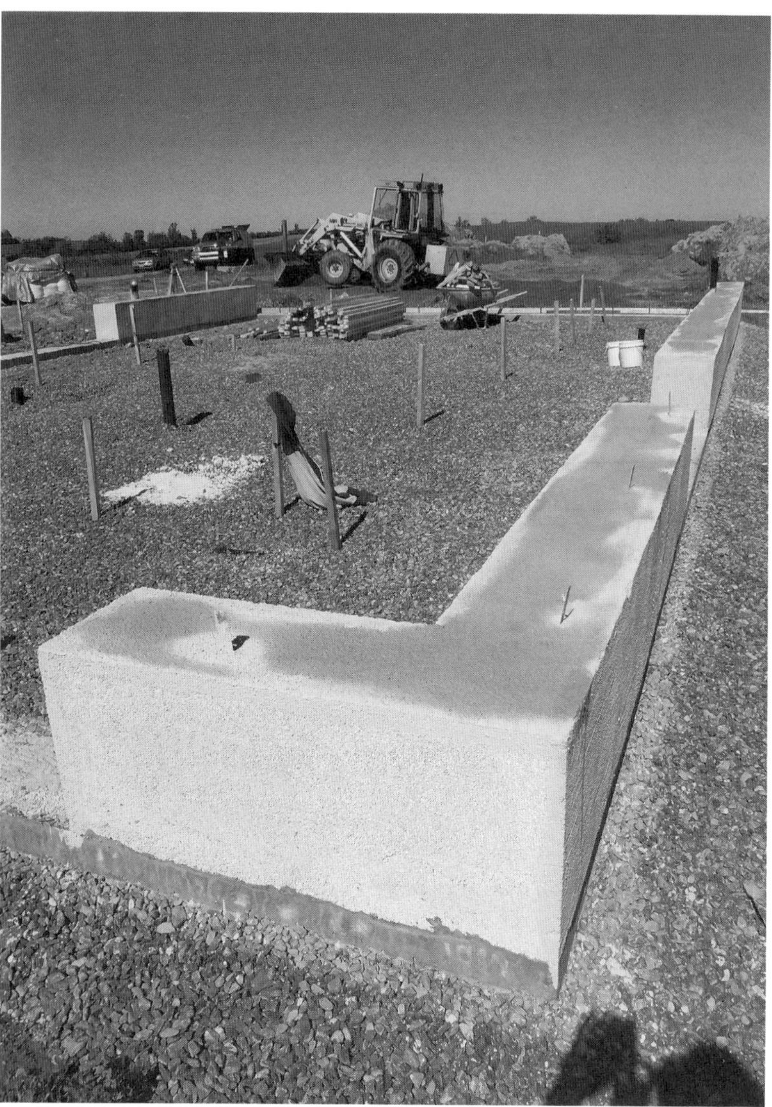

Lightweight expanded glass aggregate from Poraver.
CREDIT: CHRIS MAGWOOD

Criteria Considerations

 ***Ecosystem impacts:*** Check regional practices for aggregate quarries, as impacts can vary widely. Foaming agents may have manufacturing and on-site impacts.

 ***Embodied carbon:*** Uses much less cement than regular concrete, reducing the carbon footprint per volume ratio. Autoclaved versions have higher impacts due to emissions from the heat-and-pressure curing process.

 ***Energy efficiency:*** May not require an additional insulation strategy.

 ***Indoor environment quality:*** Inert. Waterproofing materials on the exterior may contaminate soil/groundwater.

 ***Building code compliance:*** Varies by product, manufacturer, and region. Referenced standards are widely available.

 ***Material costs:*** More expensive than regular concrete, but may not require the additional labor and material costs involved in adding insulation. Raw materials for lightweight aggregate are more costly than conventional sand/gravel.

 ***Labor:*** May require licensed installer. Lightweight aggregate mixes may be formed, mixed, and poured by homeowner.

Insulated concrete forms: *Faswall, Durisol, and Nexcem*

How the system works

There are many kinds of insulated concrete forms (ICFs). The category covered here is made from cement-bonded wood fiber. The wood fiber is typically waste-stream wood, and the wood content is mineralized by removing the sugars and rendering it inert before it is bound with Portland cement into pre-cast blocks. An insulation insert, typically made from mineral wool, is placed in the core of the block. The insulation inserts can be of various thicknesses, depending on the desired insulation value. The blocks typically range from R-12 to R-28.

Blocks come in a wide variety of shapes and sizes. They are dry-stacked in running bond on a poured concrete footing. Rebar is placed horizontally between alternate courses and vertically in every core. Concrete is poured into the cores of the completed wall forms and provides the structural strength for the system.

Criteria Considerations

Ecosystem impacts: A composite of three materials, impacts for each may vary and should be researched.

Embodied carbon: High, due to volume of cement and mineral wool used.

Energy efficiency: Should not require an additional insulation strategy. Some versions will surpass code requirements.

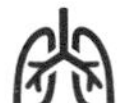

Indoor environment quality: Inert. Waterproofing materials on the exterior may contaminate soil/groundwater.

Applications	Properties
• Perimeter beams • Frost walls (including basement walls) • *Can also be used as above-grade walls*	• R-value: *1.5–2.0 per inch* • Compressive strength: *wall strength comes from concrete poured into blocks* • Density: *500–600 kg/m³ (30–38 lb/ft³)*

Waste: Block offcuts can generate a high volume of landfill waste. Finished wall cannot be dismantled and diverted.

Building code compliance: Varies by manufacturer and region. Referenced standards are widely available.

Material costs: Expensive, but eliminates the need for formwork and an additional insulation strategy.

Labor: Manufacturers recommend licensed installers. Simple dry-stacking process can be managed by homeowner.

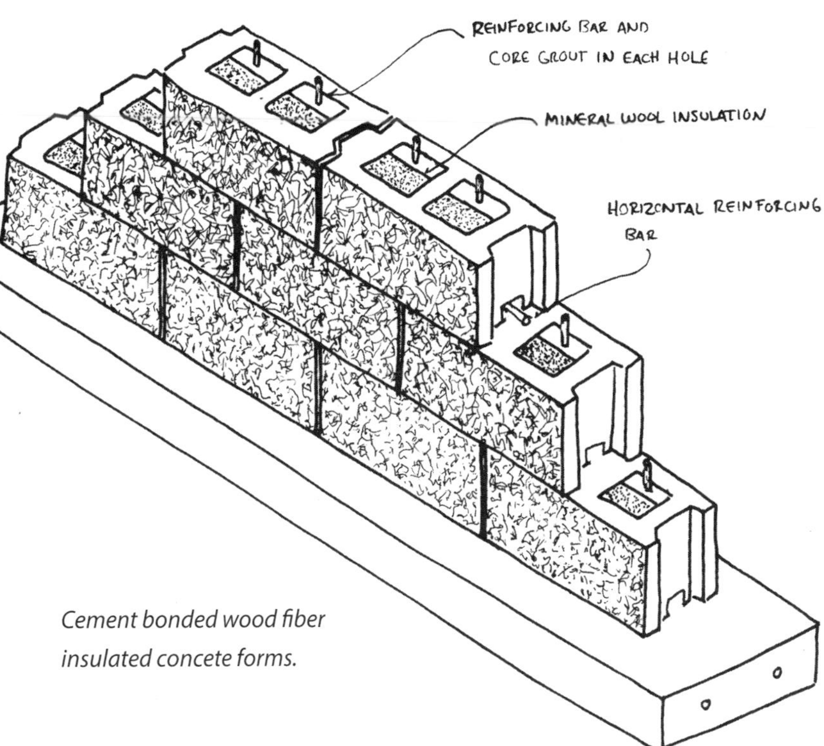

Cement bonded wood fiber insulated concete forms.

Why no foam-based ICFs?

Foam insulated concrete forms are similar to the cement-bonded wood fiber ICFs detailed above, except the forms are made from one of three kinds of foam insulation: expanded polystyrene, extruded polystyrene, or polyurethane. Though often sold as a "green" material because of their reasonable degree of energy efficiency, they are incompatible with several sustainable criteria:

Ecosystem impacts: The full "chain of custody" for foam products needs to consider the wide range of ecosystem impacts of oil exploration, extraction, shipping and pipelining, refining, and processing. Foam building products typically contain flame retardants that are very dangerous to soil, water, and the human nervous system.

Embodied carbon: Foam insulation will contribute significantly to the carbon footprint of your building.

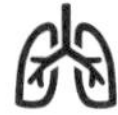

Indoor environment quality: Off-gassing of foam products is a contentious topic. Many independent researchers have noted issues of strong concern; the foam industry says there is nothing to worry about. The precautionary principle reminds us that we have often underestimated the dangerousness of chemicals we create and use that industries and regulators have deemed "safe." At the very least, the flame retardants in foam building products are known to have demonstrably ill effects on human health.

Waste: During the construction phase, the cutting and shaving of ICF blocks can generate a high volume of foam particulate that is difficult to contain and typically ends up polluting the ground (see concerns about flame retardants). At the end of life, these types of hybrid materials (foam, concrete, steel) are very difficult to separate and are likely to add substantial volume to landfills.

Resiliency: Foam products used below grade are susceptible to serious damage from ants and termites, and this damage is very difficult to notice, assess, and repair.

Though at odds with the majority in the construction industry, the author believes that the use of foam-based building materials is antithetical to the basic criteria of sustainable building, and strongly encourages you to consider other options.

Rammed earth

How the system works

A lightly moistened subsoil mix composed of clay, silt, sand, and aggregate is compressed forcefully to produce a dense and strong material. Traditionally tamped manually, much modern rammed earth is tamped with pneumatic machinery. Some type of formwork is used to contain the earth mix while it is being tamped, and the mix has an initial strength equal to the compressive force used for compaction and develops additional strength as the binder dries or cures.

There are two types of rammed earth mixtures:

- *Unstabilized* Using only naturally occurring soil ingredients. A quality unstabilized mix can have high compressive strength, but will be susceptible to degradation from exposure to water.
- *Stabilized* Using natural soil ingredients and a hydraulic cement binder. The proportion of hydraulic cement can range widely, in some cases equaling the proportion found in conventional concrete. Stabilized mixes tend to have higher strength and will be more resistant to water damage.

Rammed earth mixtures require careful formulation to ensure adequate compressive strength, integrity, and resistance to water. Research and testing should be performed with the specific ingredients to be used to determine the best mixture.

Rammed earth can be produced in several formats, all of which can be unstabilized or stabilized:

- **Formed rammed earth** — Ingredients are placed into temporary wooden forms in relatively small (4–6 inch) lifts and tamped in consecutive layers, creating a monolithic wall.

Applications	Properties
• Perimeter beams • Frost walls (including basement walls) • *Can also be used as above-grade walls*	• R-value: *0.2–0.6 per inch* • Compressive strength: *2.5–5.5 MPa (360–800 psi)* • Density: *1350–1900 kg/m³ (85–120 lb/ft³)*

Rebar or other reinforcement materials are often added to improve tensile strength and performance in seismic conditions.

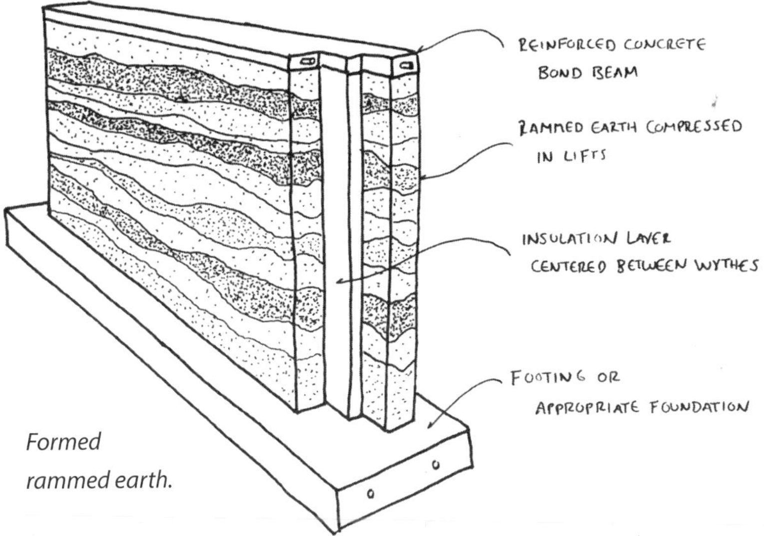

Formed rammed earth.

- **Compressed earth blocks (CEBs) and masonry units** — Ingredients are placed into a block form and manually or hydraulically compressed. The individual blocks are mortared together to create a wall.

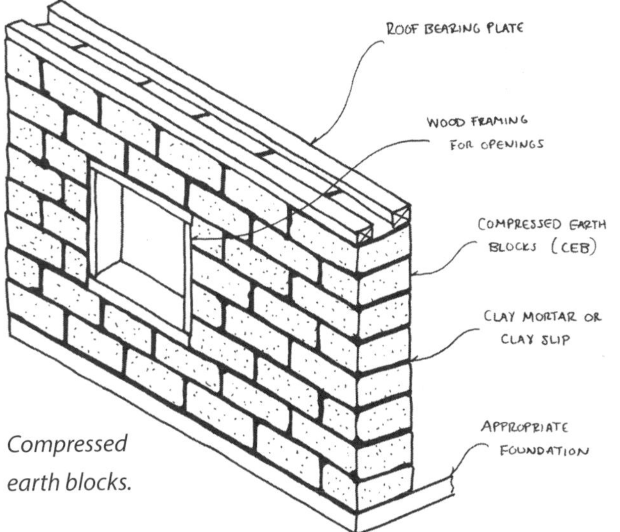

Compressed earth blocks.

- **Earthbag, or flexible form rammed earth —** Ingredients are placed into a polypropylene bag or tube, and the fabric acts as a flexible form while the mixture is being tamped, either manually or mechanically. The bag or tube typically remains in place after construction, though it is not usually functional after compression and curing. The bags can help to provide stability for mixes that are less than ideal, and they protect the mixture from erosion.

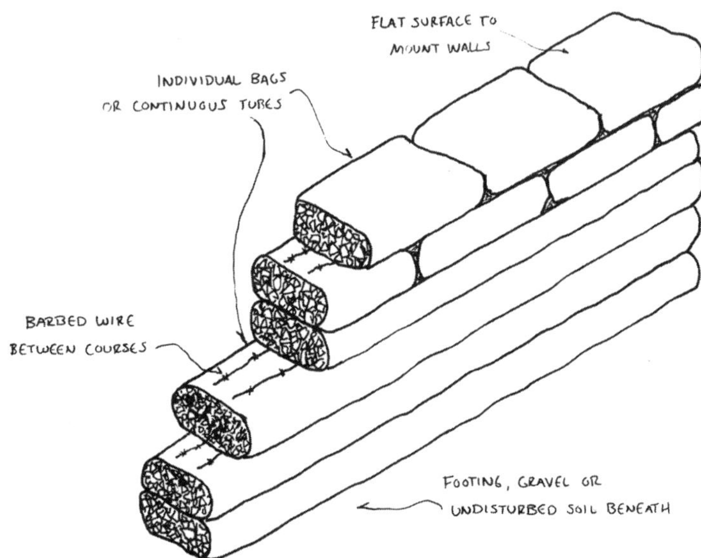

- **Earthships, or rammed earth tires —** Ingredients are placed into a used car or truck tire, and the tire acts as a form while the mixture is being tamped, either manually or mechanically. The tire remains in place as a permanent form, with tamped earth filling both the sidewalls and the open center of each tire. The tires can help to provide stability for mixes that are less than ideal, and they protect the mixture from erosion. The indentations between consecutive tires are filled with rammed earth, mortar, or other materials.

Criteria Considerations

 ***Ecosystem impacts:*** Check regional practices for aggregate quarries, as impacts can vary widely.

 ***Embodied carbon:*** Footprint will vary widely depending on the volume of Portland cement, reinforcing bar, and insulation used.

 ***Energy efficiency:*** Insulation will need to be added to most rammed earth foundations, and the full criteria implications of the insulation strategy must be carefully considered.

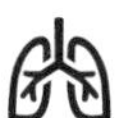 ***Indoor environment quality:*** Inert. Waterproofing materials on the exterior may contaminate soil/groundwater.

 ***Building code compliance:*** Formed rammed earth and CEB may have code references or applicable standards. Earthbag and Earthship will require alternative compliance.

 ***Material costs:*** Varies widely. Full-system costing should include formwork construction/removal, reinforcement materials, and insulation.

Labor: Formed rammed earth and CEB contractors are available regionally. Any form of rammed earth can be owner-built.

Pier and pin foundations

How the system works

A pier foundation uses a grid of posts in the ground to support loads from the building above. In some cases, piers might elevate the building completely above the ground, but piers can also anchor perimeter beams at grade level.

There are several types of pier foundation, defined by the pier material:

- **Concrete piers** — Concrete can be poured into tubular forms, or CMUs can be stacked and core-filled to create columns. The ground must be excavated to allow for the placement of the pier below the depth of frost, and the piers are typically backfilled with sand or gravel to provide drainage.

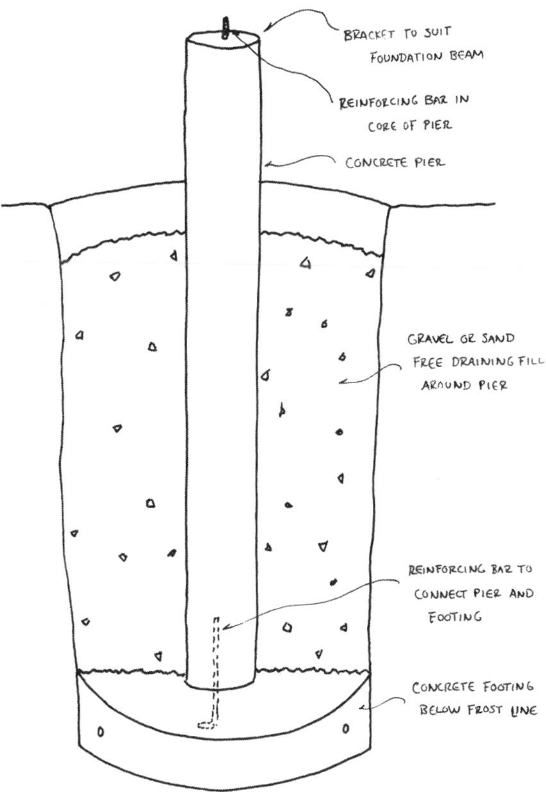

- **Helical piers** — Hollow galvanized steel tubes with a screw plate on the tip are wound into the ground with a hydraulic device until a predetermined torque reading is reached, indicating that the intended bearing capacity of the pier has been reached. Metal plates on the top of the pier are used to connect to the beam, slab, or floor framing. No excavation is typically required. They cannot be used on solid rock or in soils with large rocks.

Applications	Properties
• Piers • Support for perimeter beams • Support for slab foundations	• R-value: *N/A* • Compressive strength: *Varies widely depending on materials*

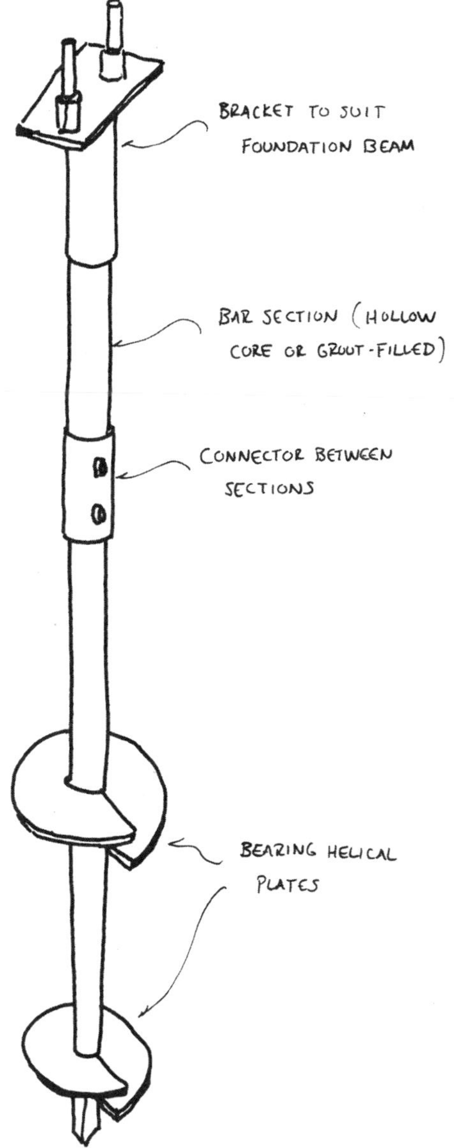

- **Pin piles** — Four steel pins are driven into the soil through a pre-cast concrete pile head. Each pair of pins acts in unison and bears on the wedge of soil immediately below. The four pins create a double-bearing wedge that is very stable and resists compressive and uplift loads. The pre-cast pier blocks and the pin diameters and lengths are engineered to meet site soil and loading conditions. Pin foundations do not require any site excavation. They cannot be used on solid rock or in soils with large rocks.

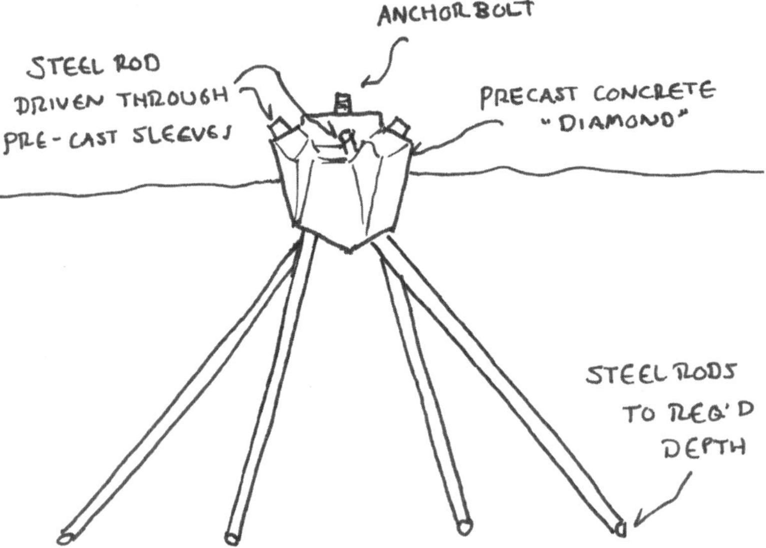

- **Wooden piers** — Wooden members of an appropriate diameter are used to support a floor deck above grade. A hole is drilled or excavated to a depth below the frost line and a gravel base and/or concrete footing are placed to receive the pier, which is made from a rot-resistant wood or treated to prevent decay. The excavation around the wooden pier is then backfilled with a well-draining sand/gravel mixture, and tamped in lifts to locate and secure the pier. Attachments between pier and beam can range from traditional timber-framing joinery to notch-and-bolt systems, to welded metal plates.

Criteria Considerations

 ***Ecosystem impacts:*** Little or no site disturbance. Metal products and concrete should be researched to determine impacts.

 ***Embodied carbon:*** Low material quantities result in relatively small footprint, even for metal and concrete options.

 ***Energy efficiency:*** Foundation does not require insulation. Floor system must be insulated appropriately.

 ***Building code compliance:*** Pier foundations are addressed in most codes. Proprietary pier systems typically provide code-compliant engineering.

 ***Material costs:*** Varies widely. Full-system costing should include any required excavation.

 ***Labor:*** Proprietary pier systems may require licensed installers. Many piers can be installed by owner-builders.

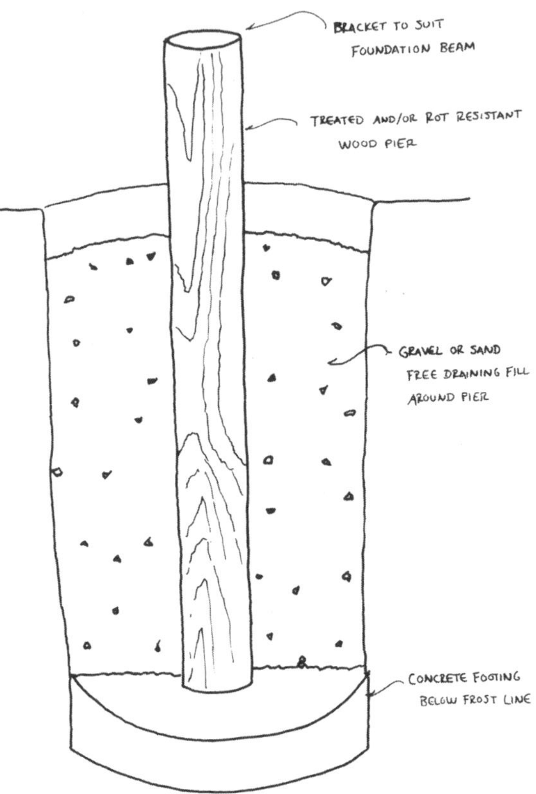

Treated wood foundations

How the system works

Foundation walls are built in the same manner as above-grade wood-framed walls (see page 122). Vertical studs are placed at regular intervals (commonly 12, 16, or 24 inches) between a single horizontal sill plate and a doubled horizontal top plate. The frame is sheathed with a structural grade of plywood on the exterior side. The wood materials are treated to prevent rot and allow the foundation to last a reasonable length of time in subgrade conditions.

There are numerous chemical-based treatments for lumber. Alkaline copper quaternary (ACQ), borates, and copper azole (CA) are all commonly available and used in treated wood foundations. The toxicity of each of these chemical treatments should be investigated to ensure that you are comfortable with using them.

Several new treatment methods — including heat treatment and silica impregnation — promise similar performance without toxicity concerns. These products are not yet widely available, but they offer the potential for healthy and feasible wood foundations.

Criteria Considerations

Ecosystem impacts: Chemical-based treatments may have significant impacts. Third-party verification for wood harvesting and production can be used to ensure minimized impacts.

Applications	Properties
• Perimeter beams • Frost walls (including basement walls)	• R-value: *Varies based on framing depth and insulation material* • Compressive strength: *Varies widely depending on assembly and materials*

Embodied carbon: Sequesters more carbon than is produced.

Energy efficiency:Insulation will need to be added to treated wood foundations, and the full criteria implications of the insulation strategy must be carefully considered.

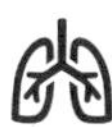

Indoor environment quality: Moisture issues may lead to mold. Waterproofing materials on the exterior may contaminate soil/groundwater.

Waste: Offcuts from treated wood may be toxic and cannot be recycled/repurposed.

Building code compliant: Compliant in most jurisdictions.

Material costs: Full-system costing must include exterior waterproofing, insulation, and interior cladding.

Labor: Contractors widely available. Can be owner-built.

Rubble trench foundations

You may have come across mention of rubble trench foundations as a low-impact alternative to conventional foundations.

A rubble trench is not really a foundation type; it is a way of connecting a foundation wall to stable, frost-free ground. With this technique, a trench is dug to the depth of the frost line around the perimeter of the building, but rather than building a wall (concrete, earthbag, tire, ICF, CMU, or AAC) from the base of the trench up to grade level, the trench is filled with compacted, well-draining stone with a continuous weeping tile at the base. The crushed stone will typically have lower environmental impacts than the materials used to build a solid wall, or at the very least be less labor-intensive than building a full wall. Yet it will offer more than adequate compressive strength to transfer building loads to the frost-free ground at the bottom of the trench.

However, it is not prudent or practical to put walls directly on the stone of the rubble trench because the rubble can only come up to grade level and cannot provide a flat, stable, airtight base on which to anchor a wall. It is not truly a foundation until another material is placed on the trench. This means that a rubble trench is always used in conjunction with another foundation system, typically a perimeter beam made from any of the materials discussed in this book. A rubble trench cannot be used to create habitable space below grade, so is not applicable for basement foundations.

Criteria Considerations

 ***Ecosystem impacts:*** Check regional practices for aggregate quarries, as impacts can vary widely. Consider the use of recycled aggregate or urbanite to mitigate impacts.

 ***Embodied carbon:*** No carbon intensive binders are used.

 ***Energy efficiency:*** Cannot be used as basement walls. Typically uninsulated.

 ***Building code compliance:*** Requires alternative compliance in most jurisdictions.

 ***Material costs:*** Aggregate costs and delivery will vary widely by region.

 ***Labor:*** Can be built by contractor or owner.

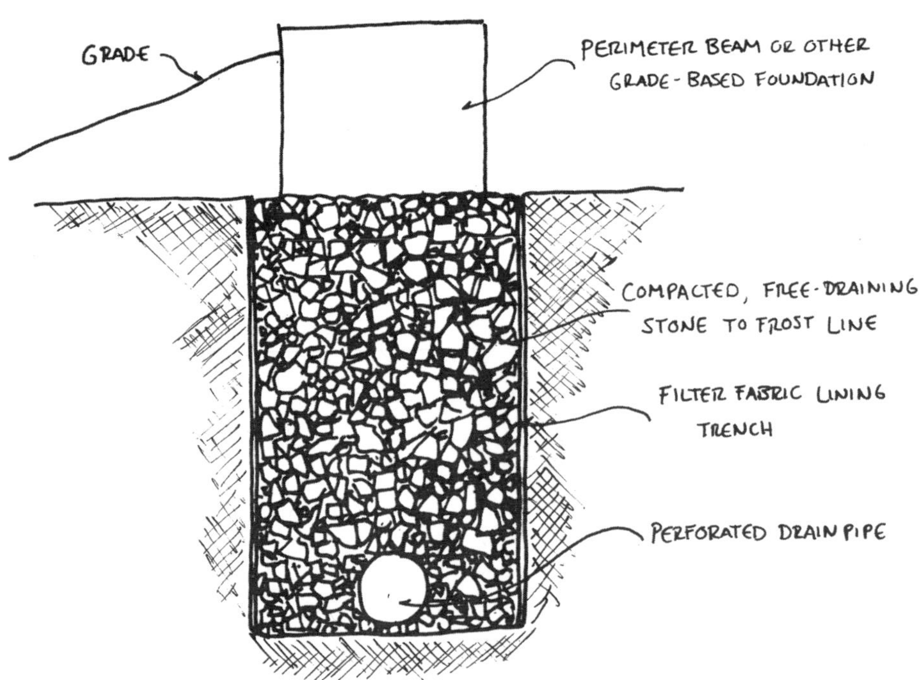

Structural Materials: Walls

Walls serve multiple functions in a building: they are key elements of the structure; they support the windows and doorways that physically define a space; they are a very important part of the building's energy efficiency; and they dominate the visual aesthetics. It is no wonder that many people come to define their home by the kind of walls it has.

A methodical comparison of walls is made difficult by the multiple functions of the wall assembly and the variety of ways in which structural, insulation, and finishing components can be arranged. For the purposes of this book, we will consider wall systems in three categories:

1. **Structural systems:** Structural walls can be skeleton frames or monolithic systems. Both types of structural walls may require insulation and/or cladding to complete the system.
2. **Insulation:** Insulation provides thermal control and may be located within a wall structure and/or on the exterior or interior side of the wall. The insulation may provide a structural role, but often does not.
3. **Cladding:** Cladding materials provide a water control layer and aesthetic finishing role on the exterior side of a wall and a finishing role on the interior.

These varying roles and the range of possible material choices require you to "mix and match" appropriate materials to create a complete assembly. There are many possible combinations that may work together as a system to suit your criteria.

Structural Wall Systems

Wood framing

Applications
• Load-bearing exterior walls • Load-bearing and partition interior walls

How the system works

Wood framing is the most common type of structural wall in North American residential construction. There are two categories of wood framing:

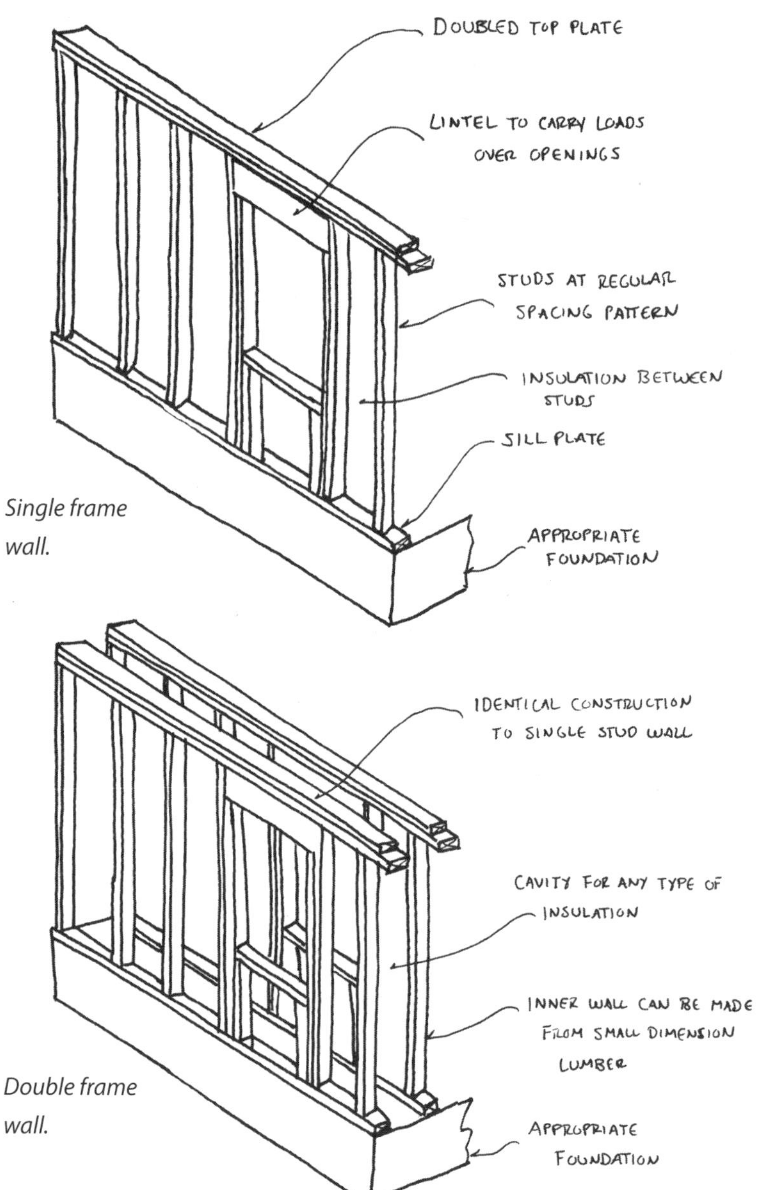

Single frame wall.

Double frame wall.

- **Stud framing** — A system of regularly spaced wooden posts or "studs" supported by a horizontal sill plate and topped with a single or double horizontal top plate to create a lightweight wall frame. Studs are typically placed at 16- or 24-inch intervals and made from 2×4 or 2×6 lumber. Insulation is placed within the framing, and (increasingly) on the exterior of the frame. Frame walls can be built with unconventional spacing to accommodate non-standard insulation types, such as straw bale. Openings in the wall are built using lintel beams to transfer loads to doubled studs on either side of the opening.

  Double stud walls are becoming more common to create extra thickness for additional insulation and avoid thermal bridges. A double stud frame with a 2×4 exterior wall and 2×3 interior wall can use a similar amount of wood as a single 2×6 wall while offering superior thermal performance.
- **Timber framing or post-and-beam** — Solid timbers of large dimensions (usually 6×6 or greater) are used to create wall, floor joist, and/or roof truss systems. The use of large-dimension wood for posts and beams increases the spacing between posts, allowing for much larger open spans than stud framing does. The expected loads determine the sizing and spacing of the beams, posts, joists, and roof members. Some codes include span charts for timber frames, but it is common to require a professional designer to determine spans and bracing requirements.

  Traditional timber framing uses wood-to-wood joinery between timbers, typically including the use of wooden pegs to fasten joints. Post-and-beam systems will use metal bracketry and fasteners to connect members. These metal fasteners can be standard, off-the-shelf materials or can be custom-welded to achieve particular structural and/or aesthetic goals.

These framing systems create a skeletal structure that is wrapped and/or filled with insulation and/or cladding materials.

Criteria Considerations

 ***Ecosystem impacts:*** Third-party verification for wood harvesting and production can be used to ensure minimized impacts.

 ***Embodied carbon:*** Sequesters more carbon than is produced.

 ***Energy efficiency:*** Insulation required. The full criteria implications of the insulation strategy must be carefully considered.

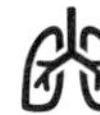 ***Indoor environment quality:*** Insulation, cladding, and finishes will determine overall IEQ. Moisture issues may lead to mold.

 ***Waste:*** Planning required to avoid high volume of offcuts.

 ***Building code compliant:*** Compliant.

 ***Material costs:*** Full-system costing must include insulation, structural sheathing, and interior and exterior cladding.

 ***Labor:*** Contractors widely available. Can be owner-built.

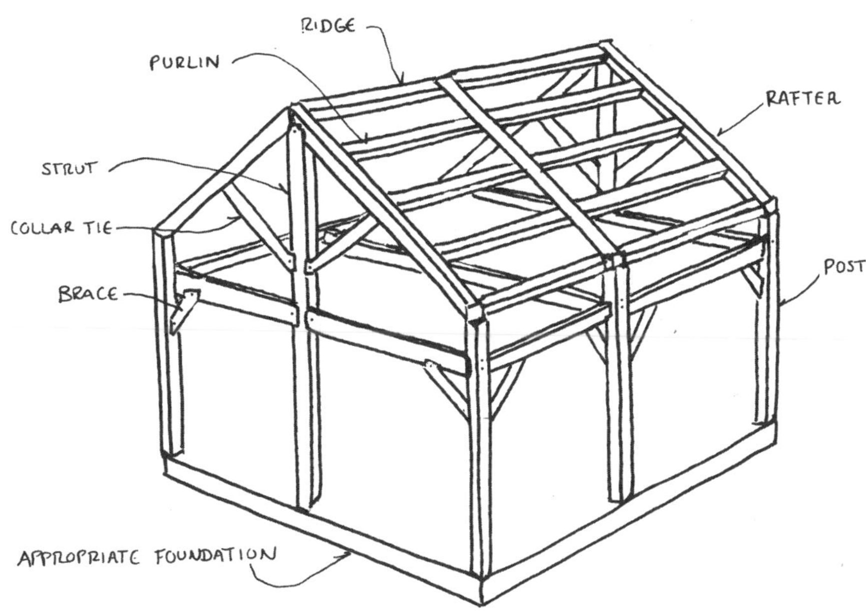

Timber frame.

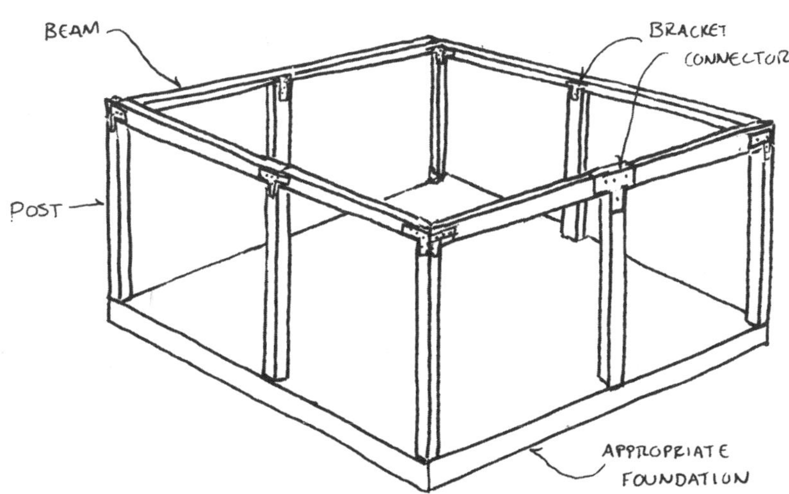

Post and beam.

Straw bale walls

Applications	Properties
• Load-bearing exterior walls • Load-bearing and partition interior walls	• R-value: *R-1.8–2.2 per inch (varies by bale density, width, and orientation)* • Compressive strength: *Suitable for structural loads in residential construction, figures vary widely depending on assembly and materials*

How the system works

Straw bale walls have been built in North America for over 100 years, but since the late 1990s this type of construction has moved from a fringe alternative to recognition within the International Residential Code (US) in 2015. The term *plastered straw bale walls* is a more accurate term for this category, as a wet-applied plaster adhered directly to the straw bales is a common feature among all types of straw bale walls. The plaster provides structural capacity, water protection, air sealing, and fire protection and is an integral element of this wall system.

The wide variety of options in straw bale construction requires you to research the best straw bale wall system to meet all your criteria. Cost, complexity, and ecological impacts can vary greatly based on framing design, plaster materials, and specific detailing. Though straw bale walls are often attractive to owner-builders who appreciate the seeming simplicity of stacking large bales to make a wall, projects will definitely benefit from having team members with straw bale construction experience; "best practices" are not as firmly established as for more conventional building types, so prior experience is a great advantage.

There are three categories of straw bale walls:

- **Infill straw bale** — Rectangular straw bales are stacked within or adjacent to a structural framing system. Frames styles include: conventional stud framing (see above) with studs spaced so one bale fits in each stud cavity; Larsen posts; post-and-beam of many variations; and traditional timber frames. The use of some type of framing system is the most common approach for straw bale walls. In some cases, the frame is designed to handle all the structural loads, but it can be much more economical and simple to design a frame that works in conjunction with the structural qualities of the plastered straw.
- **Load-bearing or Nebraska-style** — Rectangular straw bales are stacked in courses, often in running bond, to form the wall. A wooden top plate or beam at the top of the wall transfers loads into the plaster skins and provides attachment for the roof. Window

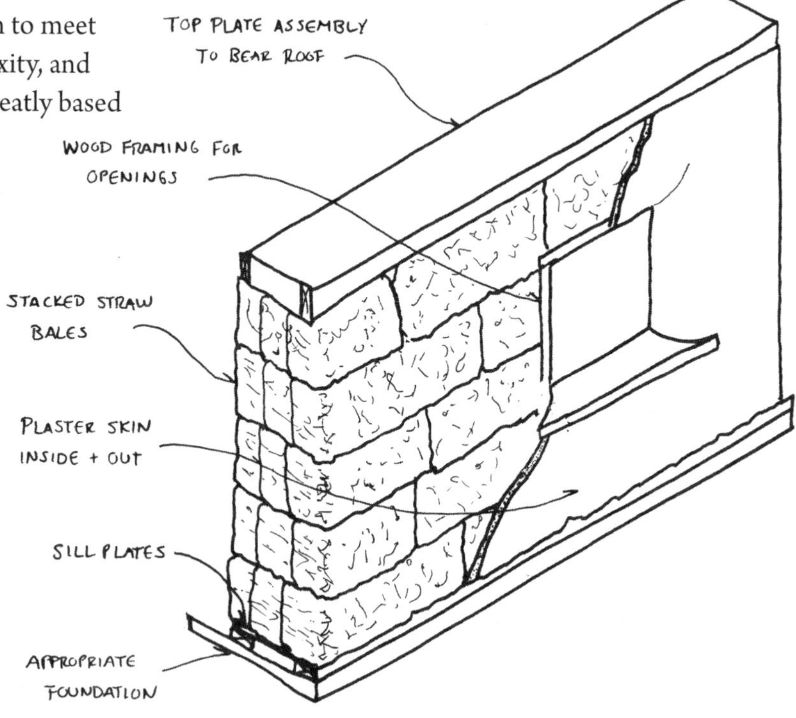

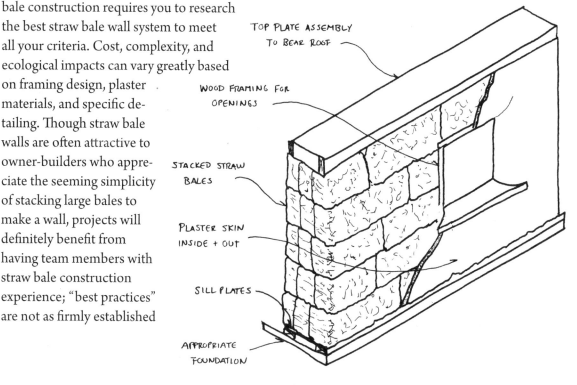

and door openings are created using wooden frames. The framing details can vary widely.

Load-bearing straw bale walls incorporate some form of pre-compression/tie-down system to connect the foundation to the roof plate to prevent uplift and to allow builders to settle and level the bale walls prior to plastering.

- **Straw bale SIPs (structural insulated panels)** — Prefabricated straw bale walls are beginning to emerge, and they capitalize on the low material costs of straw bale walls while greatly reducing the labor required for plastering, as this can be performed with the panel lying horizontal. Panels can be built on site and tipped up into place, or fabricated off site and delivered/installed via boom truck or crane.

Criteria Considerations

Ecosystem impacts: Farming practices will affect impacts.

Embodied carbon: Sequesters more carbon than is produced.

Energy efficiency: Insulation value exceeds code requirements. Airtightness strategy is important to meet high targets.

Indoor environment quality: Plaster materials and finishes will determine overall IEQ. Moisture problems may lead to mold.

Building code compliance: Building code reference in *Appendix S* of International Residential Code. (2015) Alternative compliance required in other jurisdictions.

Material costs: Full-system costing must include framing, mesh (if required), plaster materials and labor, cladding (if required), and finishes.

Labor: Some contractors available. Can be owner-built.

SIP panel guided into place and fastened.

Credit: Chris Magwood

Earthen wall systems

Applications	Properties
• Load-bearing exterior walls • Load-bearing and partition interior walls	• R-value: *0.2–0.6 per inch* • Compressive strength: *2.5–5.5 MPa (360–800 psi)* • Density: *1350–1900 kg/m³ (85–120 lb/ft³)*

How the system works

A subsoil mix composed of clay, silt, sand, and aggregate (and sometimes reinforcing fiber) is mixed to produce a dense and strong material.

There are two basic types of earthen walls:

- **Rammed earth** — A suitable soil mix is lightly moistened and compressed forcefully to produce a dense and strong material. Traditionally tamped manually, much modern rammed earth is tamped with pneumatic machinery. Some type of formwork is used to contain the earth mix while it is being tamped, and the mix has an initial strength equal to the compressive force used for compaction and develops additional strength as the binder dries or cures. Rammed earth walls can be made in two ways:
  - *Unstabilized* Using only naturally occurring soil ingredients. A quality unstabilized mix can have high compressive strength, but will be susceptible to degradation from exposure to water.
  - *Stabilized* Using natural soil ingredients and a hydraulic cement binder. The proportion of hydraulic cement can range widely, in some cases equaling the proportion found in conventional concrete. Stabilized mixes tend to have higher strength characteristics and will be more resistant to water damage.

  Rammed earth can be produced in several formats, all of which can be unstabilized or stabilized:

  - *Formed rammed earth* Ingredients are placed into temporary wooden forms in relatively small (4–6 inch) lifts and tamped in consecutive layers, creating a monolithic wall. Rebar or other reinforcement materials are often added to improve tensile strength and performance in seismic conditions.

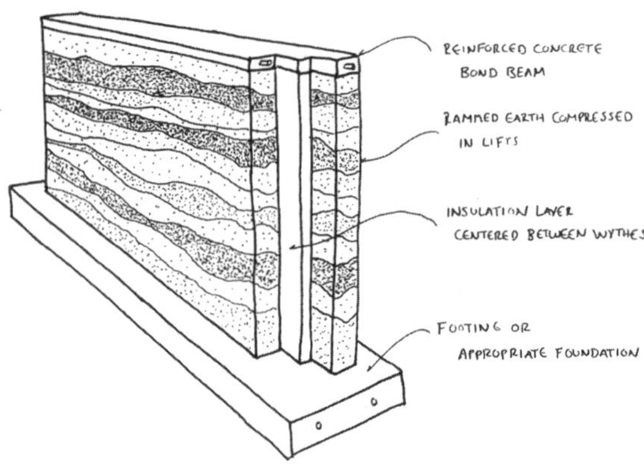

  - *Compressed earth blocks and masonry units* Ingredients are placed into a block form and manually or hydraulically compressed. The individual blocks are mortared together to create a wall.

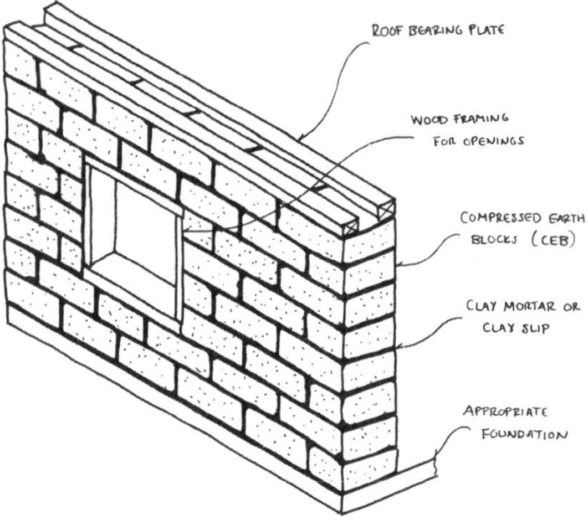

  - *Earthbag, or flexible form rammed earth* Ingredients are placed into a polypropylene bag or tube, and the fabric acts as a flexible form while the mixture is being tamped, either manually or mechanically. The bag or tube typically remains in place after construction, though it is usually not necessary after compression and curing. The bag can help to provide stability for mixes that are less than ideal, and protect the mixture from erosion.

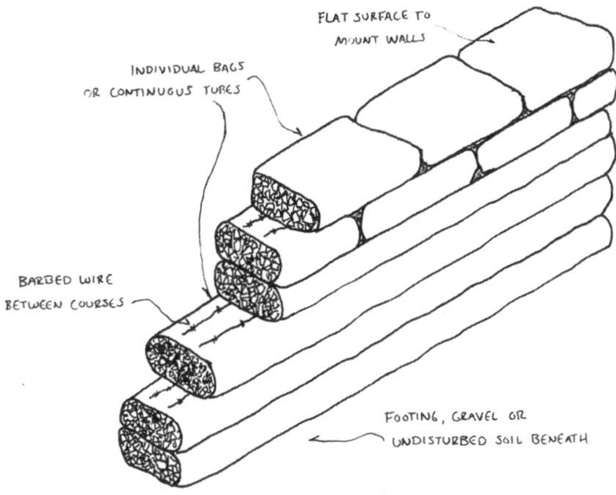

- *Earthships, or rammed earth tires* Ingredients are placed into a used car or truck tire, and the tire acts as a form while the mixture is being tamped, either manually or mechanically. The tire remains in place as a permanent form, with tamped earth filling both the sidewalls and the open center of each tire. The tires can help to provide stability for mixes that are less than ideal, and protect the mixture from erosion. The indentations between consecutive tires are filled with rammed earth, mortar, or other materials.

- **Sun-dried earth** — A suitable subsoil is wetted and mixed, most often using a natural fiber like straw, into a plastic state and is formed into a wall and allowed to sun dry to a hard state that has structural capacity. There are two main forms of sun-dried earthen walls:
  - *Adobe block (mudbrick)* A soil with clay content of 25–40% and a good distribution of sand and silt is moistened and thoroughly mixed with chopped straw or other natural fiber. The resulting mix is placed into rectangular block forms, released from the forms, and allowed to fully dry. Once dried, adobe blocks are laid up in a running bond similar to other masonry techniques. Mortar can be clay based, rather than using more conventional lime or cement mixes.

    Window and door openings use wooden, concrete, or steel lintels. At the top of the wall, a wooden, concrete, or steel beam system is used to provide rigidity and a fastening point for the roof.

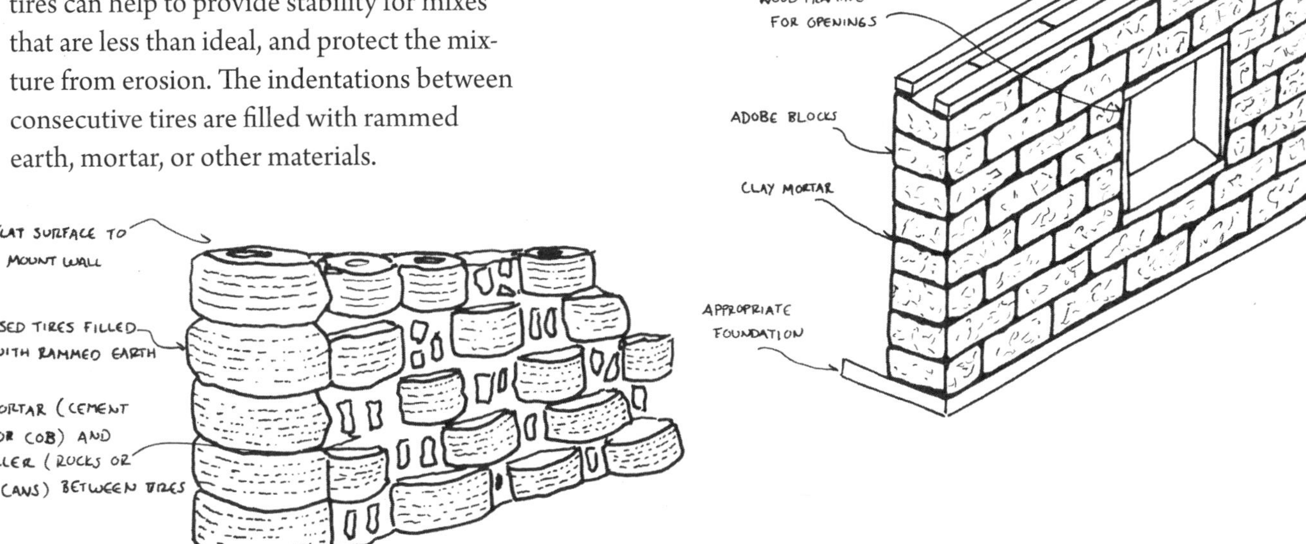

- *Cob* A soil with clay content of 10–40% and a good distribution of sand and silt is moistened and thoroughly mixed with chopped straw or other natural fiber. The resulting mix is hand-formed into monolithic walls that can bear the weight of floors/roofs. The top of the wall typically incorporates a ring beam, usually made of wood or concrete. Window and door openings are typically created using wooden frames, often built identically to openings in frame walls. Cob can be used as an infill wall with light wood frames, post-and-beam, or timber frames.

  Cob walls are not typically made with any sort of formwork, though it is possible to use forms for a more uniform wall surface.

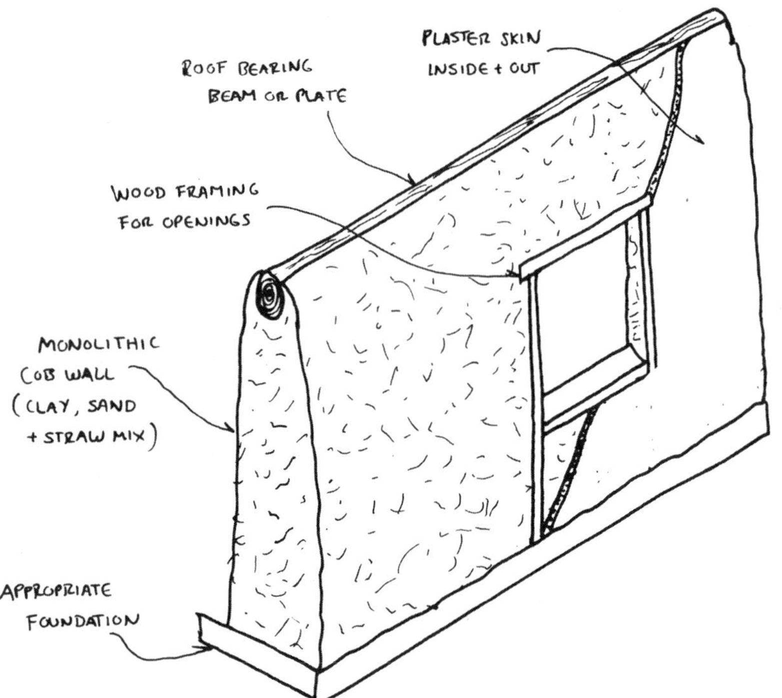

CRITERIA CONSIDERATIONS

 ***Ecosystem impacts:*** Check regional practices for aggregate quarries, as impacts can vary widely. Site-harvested or local soils can reduce impacts.

 ***Embodied carbon:*** Footprint will vary widely depending on the volume of Portland cement, reinforcing bar, and insulation used.

 ***Energy efficiency:*** Insulation will need to be added to most earthen walls. This can be achieved by adding an insulating layer on the interior or exterior side of the earthen wall, or by building double-wythe walls and insulating between the two. The full criteria implications of the insulation strategy must be carefully considered.

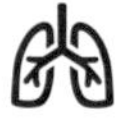 ***Indoor environment quality:*** Inert. Interior finishes will determine overall IEQ.

 ***Building code compliance:*** Formed rammed earth and CEB may have code references or applicable standards. Earthbag, earthship, and cob will require alternative compliance.

 ***Material costs:*** Varies widely. Full-system costing should include formwork construction/removal, reinforcement materials, and insulation.

 ***Labor:*** Formed rammed earth and CEB contractors available regionally. Any form of earthen wall can be owner-built.

Cordwood or stackwall systems

How the system works

Cordwood walls are formed by placing lengths of dry, rot-resistant softwood transversely across the wall and embedding the wood in a mortar (clay-, lime-, or cement-based) matrix.

There are two ways of building with cordwood:

- **Through-wall** — Mortar is used at the inner and outer edges of the wall and thermal insulation is placed in the voids between wood and mortar. In this method, the cordwood is an integral part of the thermal performance as it bridges from inside to out.
- **Double wall** — Two separate walls of cordwood are built with a continuous insulation layer between them.

Load-bearing walls use columns of cordwood wood and mortar at corners to provide stability. These are built first, and have tie-pieces that key into the walls. Many load-bearing cordwood structures are built round to take advantage of the inherent stability of round walls and avoid the need for corner supports.

Infill cordwood walls use a skeletal frame (light wood frame, post-and-beam, timber frame) and the cordwood is built between the framing members. The frame provides stability and containment for the cordwood at the corners.

Criteria Considerations

Ecosystem impacts: Wood sourcing will determine impacts, and the potential to use waste stream wood can keep impacts very low. Mortar and insulation choices will also affect impacts.

Embodied carbon: Wood will sequester more carbon than is produced. Cement- and lime-based mortar will increase carbon footprint. Insulation choice will affect overall footprint.

Applications	Properties
• Load-bearing exterior walls • Load-bearing and partition interior walls	• R-value: *1.0–2.5 per inch* • Compressive strength: *Varies widely based on mortar and wood properties*

Energy efficiency: Potential for air leakage between mortar and wood can affect performance; careful detailing is required. Overall performance determined by length of wood used and thermal properties of the wood species, as well as insulation material.

Indoor environment quality: Inert, but air leakage can negatively affect overall IEQ.

Building code compliance: Alternative compliance.

Material costs: Varies widely. Full-system costing should include cordwood preparation, insulation, and any required framing.

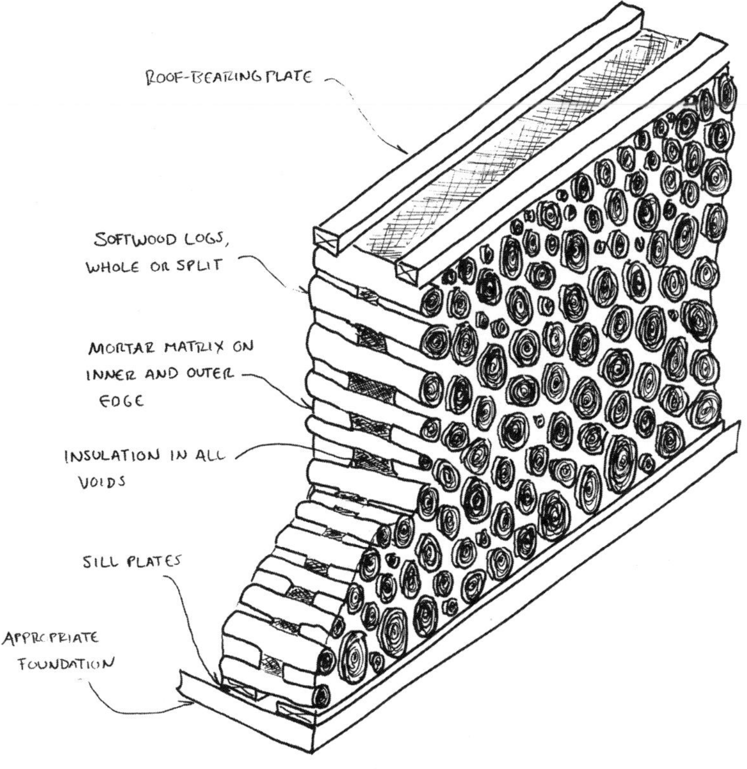

Wall Insulation

Loose-fill insulation

Applications	Properties
• Wall, ceiling, and floor cavities	• R-value: *3.6–4.0 per inch* • Density: *loose-blown ~2 lb/ft^3, dense-packed ~3.5 lb/ft^3*

How the system works

Loose-fill insulation is typically blown into cavities in floor, wall, and/or ceiling framing using a mechanical shredder/blower. On horizontal surfaces, the insulation is loose-blown and will typically settle a bit over time. In vertical wall cavities, the insulation is dense-packed using a blower with higher operating pressure to ensure the insulation will not settle.

Loose-fill insulation can be made from many different materials, including:

- **Cellulose** — Made from shredded, recycled newsprint with mineral-based fire retardant.
- **Sheep's wool** — Made from virgin or recycled natural wool with mineral-based fire retardant.
- **Fiberglass** — Made from glass fibers spun while molten.
- **Mineral wool** — Made from mineral fibers spun while molten.

Criteria Considerations

Ecosystem impacts: Natural fibers and/or recycled fibers have lower impacts.

Embodied carbon: Cellulose and wool sequester more carbon than is produced. Fiberglass and mineral wool have high embodied carbon and will contribute significantly to the carbon footprint of your building.

Energy efficiency: Good thermal properties. Loose-fill insulation fills cavities completely and prevents convective losses from gaps or voids. Thermal bridging will exist through framing members unless detailed to include insulated sheathing, double frames, or furring strips.

Indoor environment quality: Dust and microfibers introduced during installation can be problematic. Cavities should be carefully sealed, and a thorough cleaning undertaken before occupancy.

Material costs: Vary widely. Research locally available options.

Labor: Contractors widely available. Blowing equipment available to owner-builders.

Batt insulation

How the system works

Small fibers are bound together to form rectangular batts, sized to match conventional framing spacing (typically 12-, 16- or 24-inches). Batts are installed between framing members; friction against the wood keeps batts in place. Batts are cut to fit non-standard framing cavities and to conform to irregularities.

Batt insulation can be made from many different materials, including:

- **Cotton** — Recycled denim offcuts are spun into batts, often using polyester thread to help bind the fibers.
- **Cellulose** — Recycled shredded paper fibers are bound into batts, often using polyethylene terephthalate (PET) as a binder.
- **Wood fiber** — Recycled wood fibers are bound into batts, often using a wax-based binder.
- **Sheep's wool** — Recycled or virgin wool is spun into batts, sometimes using polyester thread to help bind the fibers.
- **Hemp** — Hemp fibers are spun into batts, sometimes using polyester thread to help bind the fibers.
- **Mineral wool** — Mineral fibers spun while molten are glued into batts, often using formaldehyde-based binder.
- **Fiberglass** — Glass fibers spun while molten are glued into batts, often using formaldehyde-based binder.

Applications	Properties
• Wall, ceiling and floor cavities	• R-value: *3.6–4.0 per inch* • Density: *0.75–2.0 lb/ft³ (12–32 kg/m³)*

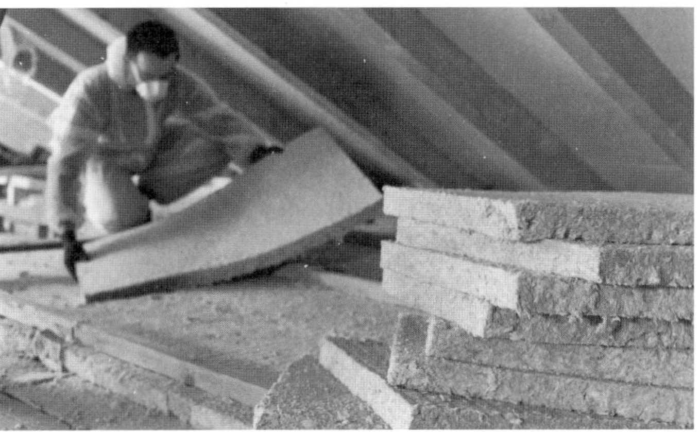

Criteria Considerations

Ecosystem impacts: Vary widely based on material. Natural fibers tend to have lesser impacts than mineral fibers. Research individual brands to determine overall impacts.

Embodied carbon: Natural fibers will sequester more carbon than is produced. Fiberglass and mineral wool have high embodied carbon and will contribute significantly to the carbon footprint of your building.

Energy efficiency: Good thermal properties. Convective losses due to gaps and voids can reduce overall effectiveness. Thermal bridging will exist through framing members unless detailed to include insulated sheathing, double frames, or furring strips.

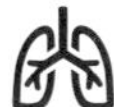

Indoor environment quality: Chemical binders can off-gas and should be avoided. Dust and microfibers introduced during installation can be problematic. Cavities should be carefully sealed, and a thorough cleaning undertaken before occupancy.

Waste: Standard batt sizing can require custom cutting and fitting and produce a large volume of offcuts that will need to be managed appropriately.

Material costs: Vary widely. Research locally available options.

Labor: Contractors widely available.

Board insulation

Applications	Properties
• Wall sheathing • Roof decks	• R-value: *2.6–4.0 per inch* • Density: *6.5–16.5 lb/ft³ (100–265 kg/m³)*

How the system works

Board insulation is produced in dimensions that correspond to typical framing dimensions, often 2×4 foot or 4×8 foot rectangles, of varying thicknesses (ranging from ½ to 4 inches). These sheets are fastened to the exterior of a wall or roof system. Some brands of board insulation come with tongue-and-groove edges to improve air sealing and thermal performance.

Board insulation can be made from a variety of raw materials:

- **Wood fiber board** — Waste wood fibers are bound together, often with a wax-based binder. Some brands may have a water-resistant coating on the panels.
- **Cork board** — Shredded cork is agglomerated using naturally occurring suberin resin and formed into boards.
- **Mineral wool board** — Mineral fibers spun while molten are glued into boards, often using formaldehyde-based binder.
- **Fiberglass board** — Glass fibers spun while molten are glued into boards, often using formaldehyde-based binder.

Criteria Considerations

Ecosystem impacts: Generally low impacts for natural and/or recycled fibers. Higher impacts for virgin mineral materials.

Embodied carbon: Natural fibers sequester more carbon than is produced. Fiberglass and mineral wool have high embodied carbon and will contribute significantly to the carbon footprint of your building.

Energy efficiency: Good thermal properties. Typically used as exterior sheathing to prevent thermal bridging when used with cavity-fill insulation materials. Can be layered to provide additional thermal performance.

Waste: Standard board sizing can require custom cutting and fitting and produce a large volume of offcuts that will need to be managed appropriately.

Material costs: Vary widely. Research specific brands available locally.

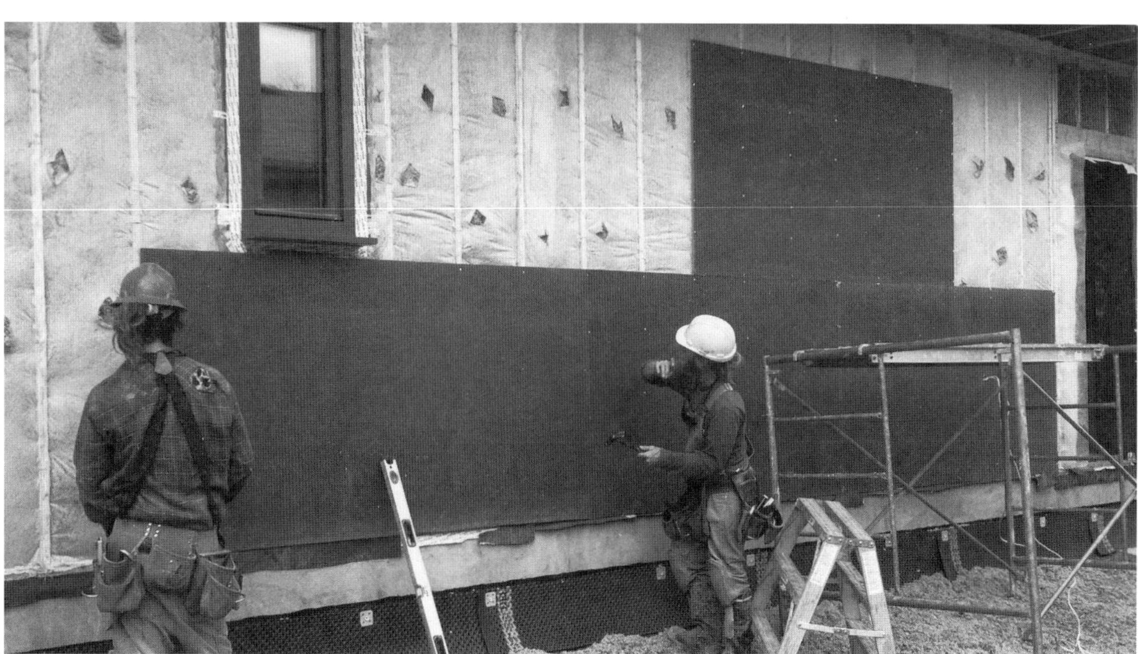

Expanded mineral insulation

Applications	Properties
• Wall, floor, and roof cavities • With hydraulic cement binder, can be used below slab floors	• R-value: *2.0–4.0 per inch* • Density: *9.5–26 lb/ft^3 (150–420 kg/m^3)* • *Grain sizes: $\frac{1}{64}$–$\frac{5}{16}$ inch (0.4–8 mm)*

How the system works

Expanded minerals are like rock-based popcorn, expanding when heated due to the expansion of trapped water or gasses. They form hard, individual grains with a high volume of air pockets when cooled. These types of insulation are completely fire-resistant and will not degrade in the presence of water. They are poured or blown into cavities.

There are four main types of expanded mineral insulation:

- **Perlite** — A naturally occurring amorphous volcanic glass.
- **Vermiculite** — A naturally occurring hydrous phyllosilicate mineral.
- **Expanded clay** — Also known as lightweight expanded clay aggregate (LECA); produced by heating clay with entrained gasses in a rotary kiln.
- **Expanded glass** — Also known as foam glass or cellular glass; produced by heating recycled glass with entrained gasses.

The European market offers expanded mineral insulation boards (perlite and cellular glass), but these are currently not commonly available in North America.

Criteria Considerations

Ecosystem impacts: Impacts vary widely depending on source minerals and harvesting practices. Recycled raw materials will typically have lower impacts.

Embodied carbon: The need for high-heat processing results in high embodied carbon levels.

Energy efficiency: Thermal properties vary widely. Loose fill insulation fills cavities completely and prevents convective losses from gaps or voids. For vertical wall cavities, material must be vibrated to induce settling and ensure cavities are full. Thermal bridging will exist through framing members unless detailed to include insulated sheathing, double frames, or furring strips.

Indoor environment quality: Inert. Dust levels can vary widely; choose low- or no-dust options if possible. Cavities should be carefully sealed, and a thorough cleaning undertaken before occupancy.

Building code compliance: Referenced standards exist for most types of expanded mineral insulation.

Material costs: Vary widely, research locally available brands for exact costing.

Labor: Few contractors. Owner-builder can install.

Bonded cellulose insulation

Applications	Properties
• Wall, floor, and roof cavities	• R-value: *1.0–2.4 per inch* • Density: *10–40 lb/ft³ (160–640 kg/m³)*

How the system works

Waste cellulose fiber from agriculture or forestry is coated in a light slurry of a clay- or lime-based binder and tamped in 4–6 inch deep lifts within temporary formwork on a wood-framed wall, roof, or floor. The forms are removed and the material is allowed to dry (3–6 weeks, typically), creating a continuous, rigid insulation.

Bonded cellulose systems are typically finished with plaster applied directly to the insulation, but they can also be clad with board or sheet materials.

There are two variations of bonded cellulose insulation:

- **Hempcrete or hemp-lime** — The lightweight core of the hemp plant, called the *hurd* or *shive,* is lightly coated with a hydraulic lime binder. The lime binder protects the insulation from fire, insects, and mold.
- **Light straw/clay (LSC)** — Hollow grain straw (from wheat, oats, barley, rice, or other cereal crop) is lightly coated with a runny clay slip. The clay protects the straw from fire and insects.

Criteria Considerations

 ***Ecosystem impacts:*** Farming practices for hemp and straw will affect impacts. Clay binder will have lower impacts than lime.

 ***Embodied carbon:*** Natural fibers sequester more carbon than is produced. Hemp sequestration will outweigh carbon footprint of lime binder.

 ***Energy efficiency:*** Moderate to good thermal properties. The ratio of binder to insulation, the density of the mixture, and the degree of compaction in the cavity will result in a wide potential range of R-values.

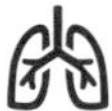 ***Indoor environment quality:*** Inert. Anti-microbial qualities of lime binder give hempcrete excellent properties. Long drying period can result in elevated humidity during construction, and drying must be complete before enclosing insulation to avoid potential mold problems.

 ***Building code compliance:*** LSC is recognized in an appendix of the International Residential Code 2015. LSC and hempcrete require alternative compliance.

 ***Material costs:*** Vary widely. Research material sources for specific costs.

 ***Labor:*** Very few contractors. Owner-builder can install.

Cementitious foam insulation

How the system works

Cementitious foam (or silicate foam) insulation is made of magnesium oxide (extracted from seawater) plus calcium (from ceramic talc) and silicate. In its wet form, the material resembles shaving cream, and it is pumped at low pressures into the area to be insulated. The material does not expand or contract after installation. It sets within an hour and dries over the course of several days. In framed walls or ceilings, one side of the frame is fully sheathed, and the insulation is applied in the stud cavity. In retrofit situations, the foam can be pumped into closed wall cavities.

Criteria Considerations

 ***Ecosystem impacts:*** Specific harvesting impacts of raw materials and processing should be researched.

Applications	Properties
• Wall, floor, and roof cavities	• R-value: *3.8–4.2 per inch* • Density: *2–5 lb/ft³ (32–80 kg/m³)*

 ***Embodied carbon:*** Low volume of raw material results in relatively low footprint.

 ***Energy efficiency:*** Excellent thermal properties.

 ***Indoor environment quality:*** Inert.

 ***Resilience:*** Material is friable (readily crumbled) and may deteriorate if exposed to repeated or significant shaking.

 ***Material costs:*** High.

 ***Labor:*** Must be installed by licensed contractor with application equipment.

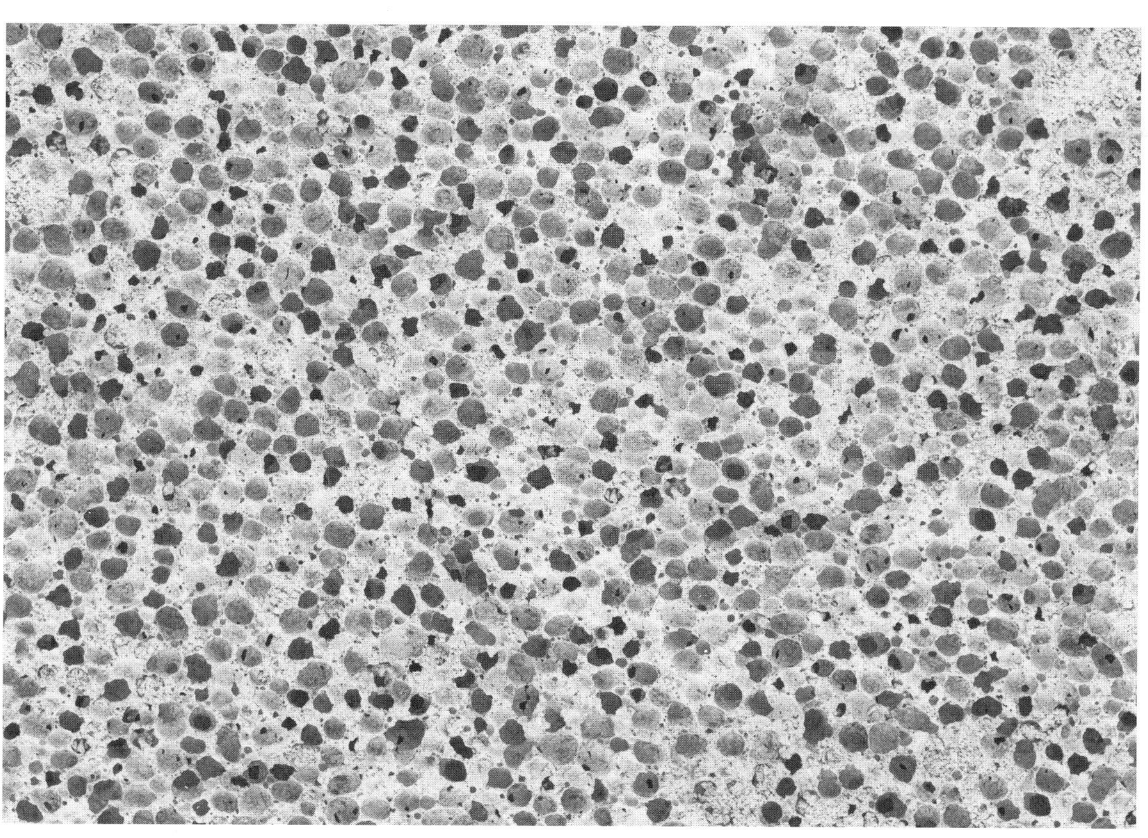

Petrochemical foam insulation

There are many types of rigid foam board and spray foam insulation. Though often sold as a "green" material because of their reasonable degree of energy efficiency, they are incompatible with many sustainable building criteria:

Ecosystem impacts: The full "chain of custody" for foam products needs to consider the wide range of ecosystem impacts of oil exploration, extraction, shipping and pipelining, refining and processing. Foam building products typically contain flame retardants that are very dangerous to the human nervous system, soil, and water.

Embodied carbon: Foam insulation will contribute significantly to the carbon footprint of your building.

Indoor environment quality: Off-gassing of foam products is a contentious topic. Many independent researchers have noted issues of strong concern; the foam industry says there is nothing to worry about. The precautionary principle reminds us that we have often underestimated the dangerousness of chemicals we create and use that industries and regulators have deemed "safe."

Waste: During the construction phase, the cutting and shaving of foam insulation can generate a high volume of foam particulate that is difficult to contain and typically ends up polluting the ground (see concerns about flame retardants). During renovations and at the end of life, spray foam is difficult to separate from other materials, and all foam products will add substantial volume to landfills.

Resilience: Foam products used in contact with the ground are susceptible to serious damage from ants and termites, and this damage is very difficult to notice, assess, and repair.

Though at odds with the majority in the construction industry, the author believes that the use of foam-based building materials is antithetical to the basic criteria of sustainable building, and strongly encourages you to consider other options.

Wall Cladding

Wood cladding

Applications

- Structural sheathing (horizontal or angled orientation, specific nailing pattern required)
- Exterior cladding (vertical or horizontal orientation)
- Interior cladding (walls and ceilings, vertical or horizontal orientation)

How the system works

Wood (typically softwood) is milled to a desired size and profile. There is a wide variety of typical wood siding profiles. Wood cladding may be fastened to wooden strapping either horizontally or vertically.

Profiles and layouts are designed to overlap or fit together at seams between planks to minimize the possibility of water penetration. Fasteners may be hidden or left exposed, depending on profile and aesthetic preference.

Options for finishing wood cladding are covered in the "Surface finishes" section, and will have significant impacts on your criteria considerations.

Criteria Considerations

Ecosystem impacts: Third-party verification for wood harvesting and production can be used to ensure minimized impacts.

Embodied carbon: Sequesters more carbon than is produced.

Indoor environment quality: Finishes will determine overall IEQ. Moisture problems may lead to mold.

Waste: Planning required to avoid high volume of offcuts.

Material costs: Full-system costing must include strapping, fasteners, and finishes.

Labor: Contractors widely available. Can be owner-built.

Plaster

Applications
• Structural sheathing • Exterior cladding • Interior cladding (walls and ceilings) • Can be used over rigid insulation or mesh/ wooden lath on framing

How the system works

All plasters are based on ratios of mineral binder and sand (and sometimes fiber) to create a mixture with the desired characteristics of stickiness and body to suit the needs of the application. Regardless of the type of mix being used, plasters require a substrate with enough "tooth" to give the plaster a good mechanical grip on the wall surface. This surface may be inherent in the wall (as with straw bale, cob, cordwood, and other natural wall systems), or it may require the use of wooden lath or some type of mesh material in the case of walls with flat surfaces.

Plaster is typically applied in multiple coats (2–3), and must dry or cure adequately before the next coat is applied. This can take anywhere from a day or two for very thin coats to several weeks for very thick coats.

There are two basic types of plaster:

- **Lime plaster** — Lime binders are produced by chemically altering limestone by applying high heat. There are three types of lime plaster:
  - *Hydrated lime* Also known as "air lime," the curing of the plaster is a chemical reaction in which the hydrated lime *carbonizes* over time. The plaster will become hard to the touch within a few days or weeks (depending on conditions) but will continue to cure and harden for decades as airborne carbon reacts with the lime.
  - *Hydraulic lime* This type of lime plaster contains fired limestone and also some amount of a pozzolanic material (finely ground powders that, in the presence of water, react chemically with lime). The pozzolan can be naturally occurring in the source limestone (*natural hydraulic lime*, or NHL) or added to the lime at the time of mixing (as with fired clay, gypsum, slag, or fly ash). In hydraulic lime plasters, there is a fast, initial setting that occurs due to the pozzolanic reaction, followed by a long-term carbonizing, as with "air limes." The amount of hydraulic reactivity varies depending on the type and quantity of pozzolan. Pozzolans may be as low as 10% and as much as 50% of the binder content.
  - *Lime-cement* Portland cement acts as the pozzolan, allowing the plaster to cure quickly in a hydraulic reaction. Lime-cement plasters typically incorporate 10–50% cement. Lime/cement mixes are commonly available as pre-formulated mortar mixes for laying brick and block.
- **Clay plaster** — There are as many different formulations for clay plaster as there are plasterers who work with the material, but all are based on ratios of clay, sand, and fiber to create a mixture with the desired characteristics of stickiness and body to suit the needs of the application.

  Clay plasters do not undergo a chemical change in the mixing or drying process, and are therefore susceptible to damage from excessive wetting that can soften and erode the mixture. These plasters are not as fragile as many believe, but unprotected clay plasters are not suitable in areas subjected to a lot of rain; in such areas, a water-resistant finish and/or roof overhang protection should be considered.

Criteria Considerations

Ecosystem impacts: Impacts vary widely based on binder type and source. Check regional practices for aggregate quarries, as impacts can vary widely.

Embodied carbon: Clay-based plaster has negligible footprint. Lime plasters can have a significant footprint depending on harvesting and manufacturing processes.

Resilience: Plaster finishes are susceptible to freeze-thaw damage in wet areas, cracking due to expansion/contraction, and chipping/breaking from impact. However, they are also relatively easy to repair and maintain without complete replacement (see "Resilience Principles" sidebar, in Chapter 3).

Material costs: Low.

Labor: Contractor availability varies regionally. Can be applied by owner-builder, but quality of finish varies by skill level.

Mineral board

Applications
• Interior cladding (walls and ceilings) • Structural sheathing • Substrate for exterior plaster

How the system works

Mineral boards come in sheet form in a variety of thicknesses (1/4, 3/8, 1/2, and 5/8 inch are common) and standard dimensions of 48 and 54 inches by 96, 120, or 144 inches to suit common wall framing dimensions.

Interior sheets are fastened to wall and/or ceiling framing with specialty screws. Interior joints are bridged with a paper or mesh tape and joint compound that is applied in several coats and sanded to create a seamless finish. Interior mineral board is commonly painted, but can also be a substrate for finish plasters and tile.

Exterior products are for sheathing only, and the joints are typically not covered; these materials are used beneath some type of cladding.

There are three types of mineral board:

- **Gypsum board** — The most common form of mineral board, also known as drywall or plasterboard. The interior version is widely used in residential and commercial construction, and has a paper coating on both sides. Exterior gypsum board includes wax in the formulation to protect against moisture and often has a fiberglass coating.
- **Cement board** — Used most commonly as a tile backer and exterior sheathing.
- **Magnesium oxide board** — Also known as "mag board." Used as a tile backer and exterior sheathing. Can also be used as an interior wall and ceiling finish and floor decking.

Criteria Considerations

Ecosystem impacts: Gypsum products tend to have lower impacts than magnesium oxide, which tend to be lower than cement board.

Embodied carbon: Gypsum products have smaller footprint than magnesium oxide, which has a smaller footprint than cement board.

Indoor environment quality: Joint compound typically contains numerous toxic substances, and the fine dust from sanding can be difficult to remove from the home. Hypoallergenic joint compound is available. Interior gypsum products can be prone to mold in moist conditions. Magnesium and cement products are inert.

Waste: Planning required to avoid high volume of offcuts.

Resilience: Interior gypsum products deteriorate rapidly in wet conditions.

Material costs: Interior gypsum products are low cost. Exterior gypsum, magnesium, and cement products are more costly.

Labor: Contractors widely available. Can be installed by owner-builder.

Clay brick and natural stone

Applications
• Exterior cladding • Exterior and interior structural wall • Interior cladding • Decorative/fire-resistant cladding

How the system works

- **Bricks** — Cast from a mixture of clay, sand, and small amounts of admixture and fired in a kiln. After firing, the clay brick is chemically altered and will not react with water.
- **Stone** — Stone of a suitable composition is harvested and either used in an unmodified form or cut/shaped to a desired dimension and profile.

Bricks and stone are laid on the foundation in successive courses, bonded by mortar. There are many different patterns, but most feature offset joints between courses. Keystone arches can be formed to create self-supporting openings, or metal reinforcement is used to form straight openings.

When used as cladding, bricks and stone are attached to structural sheathing by means of metal ties that are nailed to the wall and embedded in the mortar joint, and a space is left between the brick and the sheathing to create a rain screen. Weeper holes are left at intervals in the top and bottom courses to allow moisture to escape.

Criteria Considerations

Ecosystem impacts: Harvesting techniques will vary widely; research specific products to determine impacts.

Embodied carbon: Brick firing is a major contributor to climate change, and bricks will contribute significantly to the carbon footprint of your building. Stone footprint can vary widely depending on harvesting and processing methods.

Resilience: Very durable cladding. Repairs and maintenance relatively easy to perform.

Material costs: Vary widely.

Labor: Contractors widely available. An owner-builder may not have the required skill level.

Cement brick and stone

Many brick and stone products are actually cement-based, rather than natural clay or stone. There are many resources available to make comparisons between cement and clay brick. Cement brick manufacturers often make the claim that they are "greener" than fired clay because the process of making the cement bricks doesn't require heat. This, however, ignores the high heat input (and significantly higher carbon release) that is required to make the cement. To assess these products for your project, use data for cement/mortar and not for brick or stone.

Roofing materials as cladding

Some cladding materials are also commonly used as roofing, including:

- Metal sheets and tiles
- Cedar shake and shingle
- Slate
- Composite shingles
- Thatch

You can find more detail about each of these material options in the roofing section, below. The criteria considerations for use as cladding will match those for roofing.

Vinyl siding

Polyvinyl chloride (PVC) is a material that is included in every chemical Red List in the building industry. Though it is a popular and affordable siding, it is incompatible with many basic sustainability goals:

Ecosystem impacts: The full "chain of custody" for PVC products needs to consider the wide range of ecosystem impacts of oil exploration, extraction, shipping and pipelining, refining, and processing. Pollution emitted during PVC production is very dangerous to soil, water, and the human nervous system.

Embodied carbon: Vinyl products have high embodied energy and carbon figures.

Indoor environment quality: Though used outdoors, vinyl off-gasses throughout its lifetime (particularly when exposed to UV), and open windows and ventilation inlets near vinyl siding can introduce gasses to the indoor air.

Waste: Offcuts from vinyl siding go to landfill and are not biodegradable.

Resilience: Vinyl siding is relatively easy to damage and difficult or impossible to repair.

Though at odds with the majority in the construction industry, the author believes that the use of PVC building materials is antithetical to key basic criteria of sustainable building, and strongly encourages you to consider other options.

Roofing

The palette of practical, durable roofing materials has not changed a great deal in the past century, with rubber membranes and asphalt shingles being the newcomers (and the least environmentally friendly). There's not a long list of options to consider, and often price and availability will shrink the list to the point where you are choosing from a very limited range of materials.

Making the right choice of roofing is very important, as the roof has the most critical role in the durability of a building. A building with a weak, leaky roof will have a very short lifespan, regardless of the rest of the materials used.

More than any other element of a building, roofing requires not just the right material but also a high quality of workmanship. This factor must be weighed carefully when choosing a particular roofing material, as you will need to source the right installer or good instructions to perform the installation yourself. Warranties and workmanship guarantees are important considerations in the selection process.

An important consideration when assessing the environmental and cost impacts of a particular roofing choice is the type of strapping or sheathing that is required underneath the roofing material. Some roofing types are installed over wooden strapping, while others require a solid deck of plywood or lumber, which can significantly raise costs. Certain roof types will require specific underlayment products, and these too can add to the environmental and financial tally. Full-system costing must include all elements of the roofing system.

Metal sheets or tiles

Applications
• Roofing • Exterior wall cladding • Interior wall and ceiling cladding

How the system works

Metal roofing is a very broad category of roofing products, with many variations and styles available. The key variations between different products are:

- **Types of metal** — The two most common types of metal roofing are steel and aluminum. Copper, stainless steel, and zinc alloys are also available at premium prices. There are different environmental impacts for each type of metal, but all should be fully recyclable. Manufacturers will offer varying levels of recycled material in their roofing products, and this can have a large impact on the environmental impacts.
- **Gauge** — Different thicknesses of metal are available, typically ranging from 29 gauge at the thinnest, 26 gauge in the mid-range, and 24 gauge at the thickest. In some jurisdictions, codes may have a minimum gauge requirement. Cost rises with thickness, as does weight and difficulty of installation, with the tradeoff being increased durability.
- **Coatings** — Paints, powder coatings, alloy coatings, and stone chip coatings are among the different options available. From an environmental point of view, the coatings on metal roofing are often the least environmentally friendly element of the roof. In particular, chemical compounds including Teflon, polyurethanes, acrylics, and paints can contain elements that are environmentally detrimental during their production and are spread from the roof to the ground as the coatings wear. It is very important to do thorough research on the coatings being used on any metal roofing you are considering purchasing, especially if there is an intention to collect and use rainwater harvested from the metal roof.
- **Fastening systems** — Metal roofs use either exposed fasteners (usually screws with an integral rubber or neoprene washer) or hidden fasteners (standing seam). Exposed fasteners tend to be less expensive and easier to install, but may have a shorter lifespan.

Criteria Considerations

 ***Ecosystem impacts:*** Impacts vary by type of metal and production facility. High recycled content can offset impacts.

 ***Embodied carbon:*** High.

 ***Waste:*** Recyclable.

 ***Resilience:*** Durable and repairable. Rainwater may be harvested from roof surface.

 ***Material costs:*** Vary widely. May not require solid decking on roof, which can lower system costs.

 ***Labor:*** Contractors widely available. Owner-builder should carefully research installation guidelines.

Cedar shake and shingle

Applications
• Roofing for roofs with a minimum pitch of 3:12 • Exterior wall cladding

How the system works

Shingles are milled (saw cut), while shakes are split from the log. Many of the products called shakes today are saw milled, but done so at varying thicknesses to reproduce the more random thicknesses associated with old hand-split shakes.

Regardless of whether shakes or shingles are used, they are installed on strapping side-by-side (with a small gap to allow for expansion when wet), in horizontal courses, with successive courses set to reveal a predetermined length of the course below. The seams between shingles are offset between courses, and a cross section of the roof would show three layers of shingle at any point in the roof. The shingles come in varying widths, making it easy to stagger the joints between courses.

Shingles can be installed evenly, or staggered to create a variegated pattern. Sometimes shingles are cut into shapes on the exposed edge, creating patterns across a single course or over multiple courses.

Criteria Considerations

 ***Ecosystem impacts:*** Third-party verification for wood harvesting and production can be used to ensure minimized impacts.

 ***Embodied carbon:*** Sequesters more carbon than is produced.

 ***Resilience:*** Easy to repair. Rainwater will be contaminated by tannins from cedar and may not be usable.

 ***Material costs:*** Moderate to high.

 ***Labor input:*** Contractor availability varies regionally. Owner-builder should carefully research installation guidelines.

Mineral tiles

Applications
• Roofing for roofs with a minimum pitch of 3:12 • Exterior wall cladding

How the system works

Mineral-based tiles are installed on strapping or a solid roof deck, side-by-side in horizontal courses, with successive courses set to reveal a predetermined amount of the course below. The seams between tiles are offset between courses, and a cross section of the roof would show three layers of tile at any point in the roof.

Slate

Clay tile.

Tiles can be installed evenly or staggered to create a variegated pattern. Sometimes tiles are cut into shapes on the exposed edge, creating patterns across a single course or over multiple courses.

Ridge and hip capping is typically achieved by careful overlapping of the slates to provide positive lapping and drainage. In some cases, ceramic or metal capping is mortared onto the tiles at hips and ridges. Valley flashing is usually a metal flashing with the tiles cut to leave the metal exposed in the valley.

There are three types of mineral-based roofing tiles:

- **Slate** — Natural slate is quarried to a predetermined thickness, length, and width. Holes are drilled at the top edge to accept nails. Natural slate colors include black, red, green, blue, and grey.
- **Clay tile** — Clay roofing tiles are made from fired natural clay, often with added fluorite, quartz, feldspar, and other fluxes to reduce porosity. Tiles can be formed into a wide variety of shapes and sizes. The color of the tile is dependent on the color of the clay used, with natural red being common. Tiles are mortared together.
- **Cement tile** — Cement is pre-cast into a wide variety of tile shapes and sizes. Color is impregnated into the cement in a wide range of colors. Nail holes may be cast into the tiles or tiles may be mortared together.

Criteria Considerations

 ***Ecosystem impacts:*** Impacts vary widely based on extraction practices.

 ***Embodied carbon:*** Natural slate has low embodied carbon. Clay and cement tile will contribute significantly to the carbon footprint of your building.

Waste: Offcuts may be produced in significant quantity for roofs with hips and/or valleys and must be managed appropriately.

Resilience: Durable. Can be difficult to repair. Rainwater may be harvested from roof surface.

Material costs: Vary widely, but tend to be high. Full-system costing should include the potential need for additional framing to handle the weight of the system.

Labor: Contractor availability varies regionally. Owner-builders should carefully research installation guidelines.

Composite shingles

How the system works

Each proprietary composite shingle product is different from its competitors, but they all share a similar approach. Individual shingles composed of recycled rubber and/or plastic (sometimes incorporating waste fiber materials) are nailed side-by-side in consecutive courses, with a specified amount of reveal left on each course. The shingles are installed such that a section through any point on the roof would reveal three layers of shingle.

As proprietary products, each type of shingle will come with manufacturer's instructions for installation. These instructions should be followed carefully, and should be examined in advance to understand all of the requirements for the roofing system.

Applications
• Roofing • Exterior wall cladding

Criteria Considerations

Ecosystem impacts: Vary widely. Percentage of recycled content and the source of recycled and virgin materials will affect impacts.

Embodied carbon: Varies widely based on processing emissions.

Waste: Offcuts may be produced in significant quantity for roofs with hips and/or valleys and must be managed appropriately. Waste may not be recyclable.

Resilience: Very durable. Can be difficult to repair. Rainwater may be contaminated by chemical leaching and may not be usable.

Material costs: Vary widely. Full-system costing should include the potential need for special or proprietary underlayment material.

Labor: Manufacturers may require licensed installer. Contractor availability varies regionally. Owner-builders should carefully research installation guidelines.

Green roofs or living roofs

Applications
• Roofing, typically on low-slope roofs (3:12 or less)

How the system works

There are many different green roofing systems, ranging from simple homemade versions to proprietary, pre-manufactured systems.

There are three basic categories of green roofing:

- *Extensive* These roofs use a thin (2–6 inch) layer of lightweight growing medium with a weight of 10–25 pounds per square foot (psf). These roofs are designed to have low maintenance requirements and are often not accessible. Low-growing plants with shallow root systems and an ability to tolerate drought are usually installed.
- *Intensive* These roofs use a thick (6–36 inch) layer of growing medium with a weight of 80–150 psf. The roofs are usually designed to have regular foot traffic access. Plantings can include grasses, flowers, vegetables, perennials, shrubs, and small trees. Pathways, benches, etc., can also be designed into the roof.
- *Semi-intensive* As the name indicates, this type of green roof is designed to have weights, plants, and uses that lie somewhere between the extensive and intensive.

Regardless of the type of green roof, the basic elements are quite similar:

- A roof structure designed to handle the expected loads of the green roof; this can require substantially larger roof members than for other roofing.
- A low pitch is common for green roofs, though it is possible to build green roofs on steeper

pitches by including retaining systems to prevent growing medium from sliding off the roof when wet and heavy.

- Solid decking is used to help support the weight; depending on the type of green roof, the decking may need to be doubled.
- A waterproof membrane made from synthetic rubber compound (EPDM [ethylene propylene diene monomer], HDPE [high density polyethylene], PVC [polyvinyl chloride], and butyl rubber are common) is laid on the roof decking either as continuous sheets or as large sections that are glued or welded together.
- A drainage membrane and/or a root barrier membrane (some products combine the two in one layer) are used to provide a means for excess water to leave the roof without the growing medium getting waterlogged, and to prevent tap roots from growing down and puncturing the waterproofing membrane and causing leaks.
- Growing mediums can vary widely in composition and depth, depending on the intent for the roof. These range from dirt excavated from the foundation of the building to specially graded lightweight mediums.
- Plants are chosen to meet the intent of the roof design, to work in the type of growing medium provided, and to suit the climate and irrigation levels.
- Some living roofs incorporate irrigation systems to ensure plants are kept watered.

Criteria Considerations

Ecosystem impacts: Multiple components must be assessed and will vary widely in impacts. Most waterproof membranes are made from high-impact materials.

Embodied carbon: Multiple components must be assessed and will vary widely in footprint.

Waste: Multiple components must be managed.

Resilience: Durable. Can be very difficult to repair. Rainwater will be contaminated by growing medium and may not be usable.

Material costs: Vary widely, tend to be high cost.

Labor: Manufacturers may require licensed installer. Contractor availability varies regionally. Owner-builders should carefully research installation guidelines.

Thatch

Applications
• Roofing, with a minimum pitch of 12:12 • Exterior wall cladding

How the system works

It may seem odd for modern builders to consider that dried reed or grass stems can provide a thoroughly water-resistant and long-lasting roof, but thatch roofs have a long and successful history across a wide range of climatic zones. Modern thatched roofs are installed in almost every region of the world, though in relatively small numbers.

The system of thatching used in many wet and/or cold climates involves fastening bundles of long and thick reeds or straw to the roof strapping in successive courses, each overlapping the preceding course. The thatch is laid at a thickness (which can range from 3–8 inches [8–20 cm]) that prevents water from working its way through the layers. Thatched roofs have very steep pitches to aid in this drainage.

Many modern installations use a fire-resistant (often fiberglass) membrane under the roof strapping to prevent the spread of a fire from inside the building to the roof.

Criteria Considerations

Ecosystem impacts: Low. Annually renewable resource.

Embodied carbon: Reeds and straw will sequester more carbon than is produced.

Resilience: Durable. Relatively easy to repair. Rainwater cannot be harvested as gutters are incompatible with thatch roofing.

Material costs: Raw material is not expensive, but limited availability can make it very costly.

Labor: Contractors very rare. Owner-builder should receive thorough training prior to attempting installation.

Asphalt shingles

Asphalt shingles can be made with an organic (paper or wood fiber) or fiberglass base, saturated with asphalt and coated with mineral granules to protect the base from UV radiation. Though it is a popular and affordable roofing, it is incompatible with many basic sustainability goals:

Ecosystem impacts: The full "chain of custody" for asphalt products needs to consider the wide range of ecosystem impacts of oil exploration, extraction, shipping and pipelining, refining, and processing. Pollution emitted during asphalt production is dangerous to soil, water, and the human nervous system.

Embodied carbon: Asphalt products have a high carbon footprint.

Waste: Offcuts from asphalt shingles go to landfill and are not biodegradable. Lifespan for this type of roofing is relatively short, and will need replacing far more frequently than other roofing types. Used shingles are commonly sent to landfill and represent a large proportion of landfill volume.

Resilience: Asphalt shingles are relatively easy to damage. Leachate from the asphalt prevents many uses for rainwater collected from the roof.

Though at odds with the majority in the construction industry, the author believes that the use of asphalt roofing is antithetical to key basic criteria of sustainable building, and strongly encourages you to consider other options.

Flooring

Flooring plays a defining visual and visceral role in a building, as we see it and touch it constantly. A wide range of options faces the homebuilder when making flooring choices, and these choices represent a substantial financial investment.

Flooring obviously receives a great deal of wear and tear, and durability is of utmost importance. Patterns of wear, aesthetic preference, and type of construction may all dictate that a home has more than one type of flooring, so the decision-making process may include multiple choices, each of which suits a particular need in a particular part of the home.

Flooring considerations are actually twofold: the flooring material itself and the type of surface finish used to seal and protect it. In some cases, the two can be considered separately, as when a single flooring product has several options for finishing. In other cases, the materials come prefinished and the choice for material and finish must be made together. However, the flooring material and the finish may have very different ratings.

Plank flooring

Applications
• Finished flooring on wooden sub-floors • Finished flooring on slab sub-floors

How the system works

Planks with a tongue-and-groove profile are interlocked and either fastened to subfloor or allowed to "float." Plank widths vary from 1½–6 inches (40–150mm) for solid wood to as wide as 12 inches (300mm) for engineered flooring.

A surface finish is added to most plank floors to increase durability. Finishes will have a major impact on indoor environment quality, and must be assessed separately from the characteristics of the flooring material.

There are many different types of plank flooring, including:

- **Hardwood** — Solid wood planks from a wide variety of species.
- **Softwood** — Solid wood planks from a wide variety of species.
- **Bamboo** — Bamboo is processed into strips or into fiber, and then pressed and bonded with adhesive to form planks.
- **Engineered wood** — A surface veneer layer of the desired type of wood is bonded to a plywood or fiberboard base.
- **Engineered cork** — A surface veneer layer of cork is bonded to a plywood or fiberboard base.
- **Engineered linoleum** — A surface veneer layer of linoleum is bonded to a plywood or fiberboard base.

Criteria Considerations

Ecosystem impacts: Vary widely by material. Laminated or engineered products will have high impacts for manufacturing of adhesives. Surface finishes may have high impacts.

Embodied carbon: Natural fiber materials (wood, bamboo, cork) sequester more carbon than is produced.

Indoor environment quality: Varies widely. Surface finishes and adhesives must be researched carefully.

Waste: Offcut volume may be high and must be managed appropriately. Engineered products may not be recyclable or compostable.

Material costs: Varies widely.

Labor: Contractors widely available. Owner-builders can install.

Tile flooring

Applications
• Finished flooring, applied on wooden sub-floors or slab floors • Wall finish, applied on mineral board substrate

How the system works

Tiles are set into a bed of adhesive or mortar on the subfloor (or wall) in a chosen pattern. Once set in the adhesive, the gaps between the tiles are filled with grout to the desired level.

There are three basic types of tile:

- **Natural stone tiles** — Cut or quarried from deposits of a stone that is suitable for flooring, they may be left relatively rough and matte in finish, honed perfectly smooth, or polished to a high gloss.
- **Clay and porcelain tiles** — Made from clay, aggregates, pigments or dyes, and sometimes mineral fluxes fired together at a high temperature. The higher the temperature of the firing, the harder and less porous the finished tile. Many tiles are then glazed using mineral and/or petrochemical glazes, which are fired onto the surface of the tile during a second firing in the kiln.
- **Terrazzo and cement tiles** — Made by mixing a cement-based binder with aggregate in a form. Terrazzo is honed and polished to expose a pattern of aggregate and binder in the finished surface.

Criteria Considerations

Ecosystem impacts: Vary widely based on raw material sourcing and manufacturing process. Natural stone may have low impacts.

Embodied carbon: Natural stone may have a small footprint. Clay, porcelain, terrazzo, and cement will contribute significantly to the carbon footprint of your building.

Indoor environment quality: Inert. Adhesive, mortar, grout, and sealants may contain toxic ingredients and should be researched carefully.

Waste: Significant offcuts will need to be managed appropriately.

Material costs: Varies widely. Full-system costing should include adhesive/mortar, grout, and sealants.

Labor: Contractors widely available. Owner-builder should receive some training prior to attempting installation.

Sheet flooring

Applications
• Finished flooring, applied on wooden sub-floors or slab floors

How the system works

Rolls of sheet flooring are measured, cut, and laid in place using an adhesive.

There are two types of sheet flooring:

- **Natural linoleum** — Linoleum is formed by blending oxidized linseed oil and pine resin, into which powered limestone, wood, and cork are mixed, along with any desired pigment. This mixture is added over a wide-woven natural fabric, such as jute or burlap backer. The linoleum is then cured in a kiln or drying room for a number of days to complete the polymerization of the oils and make the mixture hard.
- **Cork** — The bark of the cork oak tree is harvested (in cycles of 7–12 years) and air-dried into sheets. Wine corks are punched from the sheets, and the remainder is ground, boiled, and formed with adhesives into sheets.

Criteria Considerations

Ecosystem impacts: Cork and linoleum are based on renewable resources with low impacts.

Embodied carbon: Natural fiber materials sequester more carbon than is produced.

Indoor environment quality: Inert. Adhesives and surface finishes may contain toxins and should be researched carefully.

Material costs: Can be high. Full-system costing should include adhesives.

Labor: Contractor availability varies regionally. May require special tools or training to install.

Concrete flooring

Applications
• Floor slabs. Can be a skim-coat finish or a slab that is serving a structural role.

How the system works

A mixture of Portland cement and aggregate is poured to the desired thickness over a stable base able to support the weight of the material. Concrete can be poured with pigment mixed in, or pigment may be cast onto the wet surface of the slab and worked into the concrete during finishing. Concrete can also be poured and finished with no color, and have a color finish applied to the surface after the slab is cured. Concrete can be finished to a wide variety of surfaces, from roughly textured to highly polished. Concrete surfaces can also be stamped with a wide variety of patterns.

There are two types of concrete floor:

- **Slab floor** — A structural element in the building with the surface used as a finished floor.
- **Concrete skim floor** — Applied over an existing sub floor in a relatively thin layer to provide a finished floor surface.

Criteria Considerations

Ecosystem impacts: Check regional practices for aggregate quarries, as impacts can vary widely. Consider the use of recycled aggregate to mitigate impacts.

Embodied carbon: Cement production is a major contributor to climate change, and concrete will contribute significantly to the carbon footprint of your building.

Indoor environment quality: Inert. Pigments and surface finishes may contain toxins and should be researched carefully.

Material costs: Full-system costing should include reinforcement materials and finishes.

Labor: Contractors and suppliers widely available. Can be formed, mixed, and poured by owner-builder.

Vinyl flooring

Many stick-down tile or sheet flooring products called "linoleum" are actually made from polyvinyl chloride (PVC), a material that is included in every chemical Red List in the building industry. Though it is popular and affordable flooring, it is incompatible with many basic sustainability goals:

Ecosystem impacts: The full "chain of custody" for vinyl products needs to address the wide range of ecosystem impacts of oil exploration, extraction, shipping and pipelining, refining, and processing. Vinyl production pollution is very dangerous to soil, water, and the human nervous system.

Embodied carbon: Vinyl products have very high embodied energy and carbon figures.

Indoor environment quality: Vinyl off-gasses throughout its lifetime (particularly when exposed to UV) and contributes to poor indoor air quality. Adhesives may also be toxic.

Waste: Offcuts from vinyl siding go to landfill and are not biodegradable.

Resilience: Vinyl flooring is relatively easy to damage and difficult or impossible to repair.

Though at odds with the majority in the construction industry, the author believes that the use of PVC flooring is antithetical to key basic criteria of sustainable building, and strongly encourages you to consider other options.

Surface finishes

There are many surfaces in a building that require a finish to protect the underlying material and/or add an aesthetic dimension. Even a small home may require thousands of square feet of surface treatments. It is rare that one type or one color of finish is chosen for the whole home; there are usually multiple finishing decisions that have to be made.

In many cases, the surface finish is a key element in the durability of the material it is protecting. The finish takes the brunt of exposure to the elements, wear and tear, and cleaning.

We ask a lot of our finishes, and modern science has been successful in creating finishing products that offer excellent durability, color choice and fastness, ease of application, and adhesion. Unfortunately, in the pursuit of such qualities, these products have become proprietary "chemical soups." Even the greenest petrochemical finishes rely on extraction and manufacture of chemical components that have a wide variety of health and environmental impacts throughout the entire chain of production, application, use, and disposal.

The majority of the finishes described in this chapter fit under the heading of "natural finishes." They use naturally occurring ingredients that are free of petrochemical products. They are viable on a wide range of surfaces and materials, and offer low impacts and low or no toxins — from raw material acquisition through to final application. In some cases, the final product may not offer quite the same degree of durability, color choice, ease of application, or adhesion as their petrochemical counterparts — a small tradeoff for vastly reduced health and environmental impacts. In many cases, the natural finishes offer a beauty and richness that cannot be matched by petrochemical finishes.

The one exception to the focus on all-natural finishes is the section on nontoxic latex paints. There are a few paint companies making an attempt to create actual nontoxic latex paints; while these are a vast improvement over petrochemical paints of the past, they are not entirely clean and free of petrochemicals, nor can the chain of production be guaranteed to be clean and nontoxic. However, because there is increasing interest in truly nontoxic latex paints, we include them as a category in this chapter. They will offer homeowners the same level of performance expected from commercial paints with greatly lowered impacts.

Natural paints

Applications
• Interior walls and ceilings • Exterior walls • Interior and exterior trim and wooden elements

How the system works

This category of finishes captures a wide array of materials. They share in common a mixture of some form of pigment, filler, binder, and solvent that is applied in a thin layer that dries to become a solid film.

Paint mixtures can be applied by brush, roller, sponge, and/or trowel, depending on the consistency and the desired appearance of the finish. Application techniques such as rubbing, sponging, and trowel burnishing will lend the finish different appearances, even using the same mixture.

There are commercially produced versions of all the natural paints included here, but homeowners can also mix their own from the raw materials.

There are many types of natural paint:

- **Oil** — *Suitable for interior and exterior surfaces, including plaster, drywall, wood, and some products can be used on metal.*

  Natural oil paints are those in which the resin is natural, siccative (oxidizing) oil (typically cold-pressed linseed, but can be semi-siccative oil like walnut, hemp, poppy, tung, sunflower, safflower, soya, and even fish oil). Linseed oil (pressed from flax seed) is by far the most common.

  There are two kinds of natural oil paint. *Emulsion formulas* are mixed with water to provide an interior paint that is quick-drying and barely distinguishable from conventional paint in coverage and application. *Pure oil paints* are slow to dry and require a solvent to dry sufficiently for use as paint. Natural turpentine and citrus solvent are the two most common drying agents for natural oil paint. Solvent content in oil paint can be very high, and even natural solvents contain volatile organic compounds (VOCs). During the long drying process for oil paints (though dry to the touch in a matter of days, full drying takes much longer), the oil undergoes a natural polymerization. In recent years, advances in polymerizing natural oils have resulted in oil paints that dry faster and harder without requiring as much solvent. Some of these processes are natural, and others are chemical.
- **Lime** — *Suitable for interior or exterior surfaces, including plaster and drywall.*

  Powdered, fired limestone is mixed with water and applied to a surface, where it chemically re-carbonizes into a durable material that resembles the original stone. The lime will reach its "working strength" in 30 days, and continue to strengthen over time. Multiple coats require curing time between applications of at least 24 hours to allow time for carbonization to begin. For thicker coats, longer curing time between coats is recommended.

  Simple lime washes combine just powdered lime and water. The addition of aggregate, fillers, and/or fiber in the mix gives lime paint enough body to create thicker coats. Pigment can be added to wash or paint. Lime paint can fully cover a substrate, filling small cracks and pores and surface irregularities.

  Lime is stable when exposed to water and is naturally anti-microbial.
- **Clay** — *Suitable for interior surfaces in dry areas, including plaster and drywall.*

  Clay paints are as old as human civilization, as they can be made with clay that is naturally occurring in most regions. Variation in the type, size, and quantity of clay and aggregate can result in finishes ranging from grainy, heavily textured surfaces to polished, glass-like surfaces; the finer the aggregate, the smoother the texture.

Once applied, the mixture dries and hardens. There is no chemical change in clay paint, so exposure to water can soften or erode the paint. In areas where high wear or water exposure is expected, an oil or wax finish can be applied over the clay paint to add a further degree of protection.

- **Milk or casein** — *Suitable for interior or well-protected exterior surfaces, including wood, plaster, and drywall.*

  The effectiveness of milk paint is based on the properties of casein molecules, which contain a glue-like substance that is freed in the presence of base materials like lime or borax. Powdered fillers like clay and/or calcium carbonate give the paint body, and pigments add coloration. Microfibers (such as cellulose) can add further body, and a number of admixtures may be included to give particular properties to the paint.

  Once mixed with water, the paint requires 20–60 minutes to allow the reaction between the casein and lime to transpire.

  In areas where high wear or water exposure is expected, an oil or wax finish can be applied over the milk paint to add a further degree of protection.

- **Silicate dispersion** — *Suitable for exterior and interior surfaces over mineral substrates including plaster, concrete, and brick.*

  Silicate paints are also known as *water glass, silicate dispersion paint, silicate mineral paint,* or *silicate emulsion paint.*

  Sodium and potassium silicate are water soluble, and when dispersed onto a mineral surface such as clay, lime, or cement plaster, they will bind with silicates in the substrate, petrifying into a microcrystalline structure. These tiny pores are ideal for repelling liquid water but do not restrict the passage of vapor, giving this paint the ideal balance of moisture-handling properties. Exterior surfaces are protected against the entry of rain, but allow migrating humidity to leave the wall. Interior surfaces are washable, and the paint can be used in kitchens and bathrooms.

Criteria Considerations

Ecosystem impacts: Natural paints have much lower impacts than petrochemical paints. Impacts vary widely based on source ingredients and should be researched carefully.

Embodied carbon: Low for most products.

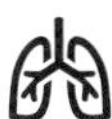

Indoor environment quality: Typically, natural paints contain only nontoxic ingredients. Oil paints may produce a high initial volume of VOCs. Pigments should be researched carefully, as some natural pigments may be toxic. Many natural paint manufacturers offer full disclosure of all ingredients.

Waste: Most natural paints are 100% biodegradable and nontoxic. Washing can be done in household sinks or outdoors.

Material costs: Range widely. Basic ingredients tend to be low cost. Manufactured paints can be similar to conventional paint costs or more expensive.

Labor: Paint contractors may not work with natural paints. Owner-builder can apply.

Acrylic (latex) paint

Applications
• Interior and exterior surfaces

How the system works

While "latex" is the common name for water-based acrylic paints, naturally derived latex (a milky substance found in certain flowering plants) is not used in the vast majority of household paint. The binder portions of the paint tend to be blends (dispersions) of acrylic (polymethyl methacrylate or PMMA), styrene-acrylic, vinyl, and polyvinyl acetate (PVA), with different proportions of those three ingredients based on the paint's intended use and cost. Fillers, which are inert bulking agents, are used to create a desired viscosity and texture.

There are different formulations of acrylic paint to suit particular applications. The higher the quality and durability of the paint, the higher the percentage of acrylic. Cheaper, less durable paints have more vinyl and polyvinyl acetate.

Due to a generally high level of environmental impacts and toxicity, this category of finishes would not be included in this book if not for the concerted efforts of a small number of acrylic paint manufacturers making serious attempts to reduce the toxic content of their products and clean up the manufacturing process. A sustainable builder should not consider using acrylic paint unless sourced from manufacturers with the highest ascertainable standards.

The truth about no-VOC paint

Low- and no-VOC designations are *not* indicators of a lack of toxins in paint. VOC regulations were not instituted to protect human health, but rather to address ozone smog issues, and as a result *only* VOCs that contribute to ozone smog are addressed in VOC regulations. Chemicals like highly toxic ammonia and acetone are not regulated, nor are toxic combinations that form when paints are released into the air. Proprietary "additives" like biocides and fungicides are not included, even when they contain formaldehyde and other VOCs.

Criteria Considerations

Ecosystem impacts: The full "chain of custody" for acrylic paint products needs to consider the wide range of ecosystem impacts of oil exploration, extraction, shipping and pipelining, refining, and processing.

Embodied carbon: High, though actual volume of paint used will not contribute excessively to the overall footprint of the building.

Indoor environment quality: Poor. Even paint that is labeled as "no VOC" does not guarantee that the product is free from toxic off-gassing (see sidebar). Many chemicals in paint may still be toxic, but don't contribute to VOCs. Research carefully before using these.

Waste: Acrylic paint residue in waterways (from cleanup in sinks) is an environmental problem, and leftover paint contributes significantly to toxic waste volume at waste facilities. Some recycling programs exist.

Material costs: Range widely. Nontoxic versions will cost more.

Labor: Contractors widely available. Owner-builder can apply.

Natural plaster

Applications
• Finish coat for application over most interior substrates, including plaster, drywall, magnesium oxide board, brick, and sheet wood materials. • Finish coat for application over some exterior substrates, including plaster, magnesium oxide board, brick, and concrete.

How the system works

Finish plasters are applied as a very thin skim coat (⅛–¼ inch [3–6 mm]). Mix ratios can vary widely, but 1 part of binder to 2–3 parts of fine aggregate is common. Admixtures are common to strengthen the plaster, and pigment can be incorporated for embedded color. Most natural pigments take very well to natural plaster, giving rich colors and a sense of depth that is unlike painted surfaces.

For application over smooth substrates, an adhesion coat is used to help the plaster to bond to the flat surface. This coat is typically a sand and glue (natural flour paste or casein glue) mixture that is brushed or rolled onto the wall to create a textured surface. The plaster is trowel applied and a wide range of textures and finishes are possible, from rough to highly polished. Sealant or silicate paint may be used to protect the finish plaster for exterior applications.

There are two types of finish plaster:

- **Lime** — Hydrated or hydraulic lime is the binder. Suitable for interior or exterior applications.
- **Clay** — Naturally occurring or refined bagged clay is the binder. Suitable for interior applications in dry areas.

Criteria Considerations

Ecosystem impacts: Low. Variable based on source of ingredients.

Embodied carbon: Low. Variable based on source of ingredients.

Indoor environment quality: Inert. Research required for potential pigment toxicity.

Material costs: Low.

Labor: Contractors not widely available. Owner-builder can learn to mix and apply.

Natural oils and waxes

Applications
• Interior wood, plaster, concrete • Interior walls, floors, countertops, window sills, furniture

How the system works

A degree of protection from water, UV, and wear can be provided by natural oil and/or wax finishes on most natural, porous materials. Formulations vary widely, as does the degree of protection.

- **Oil** — Natural, siccative (oxidizing) oil (typically cold-pressed linseed, but can be semi-siccative oil like walnut, hemp, poppy, tung, sunflower, safflower, soya, and even fish oil) can be used to protect, seal, and enhance the color of porous surfaces.

  Oils for finishes may be pure oil or a blend of oils formulated to provide desired characteristics. There are many such blends, and their intended use should be researched before committing to use of a particular product. Some oil finishes are referred to as "hard oils." This refers to a lack of solvents or thinners in the mix, and not necessarily to a more solid or durable surface.

  Oil finishes are applied in thin coats; if applied too thickly they may never completely harden. Most oil finish products are blended for best drying characteristics, or are polymerized during production to speed curing times. Multiple coats may be required, depending on the product, the substrate, and the level of protection required from the finish.
- **Wax** — Numerous plants and animals produce natural wax, but the most common are beeswax and carnauba wax. Most finishes use petrochemical-based wax, and even many products sold as natural wax includes some percentage of petrochemical content. Most wax finishes will include some form of solvent (natural or petrochemical) to thin the wax and make it easier to apply.

  Wax is rubbed onto a surface in a thin coat, allowed to start to harden, and then buffed to provide the degree of polish desired.
- **Oil wax** — Oil is sometimes blended with natural wax in a single product.

Criteria Considerations

Ecosystem impacts: Vary widely depending on the source of oil or wax.

Embodied carbon: Low.

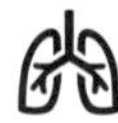

Indoor environment quality: Natural oil and wax products are nontoxic. Many products blend natural ingredients with solvents and/or chemical additives, and many of these contain toxins. Natural products may emit VOCs while curing. Research products carefully.

Material costs: Vary widely. Full-system costing should account for the number of coats required.

Labor: Few contractors work with natural oil and/or wax. Owner-builder can apply.

Natural wallpaper

Applications
• Interior wall covering

How the system works

Decorative papers are printed and/or embossed with color and pattern. Grass or fiber wall coverings are woven and/or stitched. Both are made in rolls and applied to the substrate using an adhesive. Rolls are made in regular widths and are applied side-by-side to cover a desired area.

Many conventional wallpapers and coverings are made of or have a surface treated with polyvinyl chloride (PVC), acrylic, polyurethane, or toxic inks, and they are they are adhered to walls with adhesives that are high in VOCs, fungicides, fire retardants, and other toxic ingredients. Only products verified to be nontoxic should be considered.

Criteria Considerations

Ecosystem impacts: Low. Impacts will vary based on sourcing of raw materials.

Embodied carbon: Natural fibers sequester more carbon than is produced.

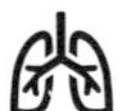
Indoor environment quality: Natural products are inert. Many products in this category contain toxic ingredients and should be avoided. Glue and paper can grow mold in moist conditions.

Material costs: Natural options tend to be expensive.

Labor: Few contractors specialize in natural options. Owner-builder can install with proper training.

Chapter 11

Mechanical Systems

THE MODERN HOME is as much about its mechanical systems as its structure. Electrical generation and distribution, water collection and distribution, wastewater removal and treatment, heating, cooling, and ventilation are all key components of any residential construction project. Though in many cases these systems are hidden, their functionality is what makes a modern home so comfortable.

These systems are also what cost homeowners and the planet so dearly. The ongoing financial and environmental costs of keeping every home serviced with our current methods is not sustainable, and the choices we make today about mechanical systems in buildings will have a huge influence on our collective sustainability and resilience in the future.

A Life Cycle Analysis (discussed in Chapter 3) of buildings show that the "operating energy" used to make our mechanical systems function is by far the largest contributor to a building's energy footprint. Our mechanical systems dictate our impact on water resources, air pollution, and greenhouse gas emissions.

There are many points along the line between our current energy-intensive practices and the no-energy option of eliminating all mechanical systems. This chapter is intended to give an overview of the options that exist and their potential for reducing impacts while maintaining comfort.

Mechanical Systems: Water

For most of human history, we have spent a huge amount of our time collecting, moving, and storing water, mostly with manual labor. There are still many places in the world where turning a tap to receive an endless supply of potable water is not the norm. And there are good reasons that we should no longer consider water to be the cheap, disposable resource we've come to expect at every faucet.

Water issues are complex. In the simplest terms, we tend to consider the source, quality, and quantity of water that comes out of taps and faucets in the home. These are obviously very important issues, and there are significant concerns on all three counts that help to direct more sustainable choices.

What we rarely consider are the high energy and carbon costs that accompany water. Figures for Ontario, Canada, show that "water and wastewater services together represent a third to a half of a municipality's total electricity consumption," and that "municipalities, largely responsible for the provision of water in Ontario, have been reported to consume more electricity than any industrial sector outside Pulp and Paper."[17] This does not take into account the energy used for private residential and commercial wells or industrial water pumping. Conserving water, when seen in this light, is about more than reducing the use of a valuable natural resource, it is also inherently about energy conservation.

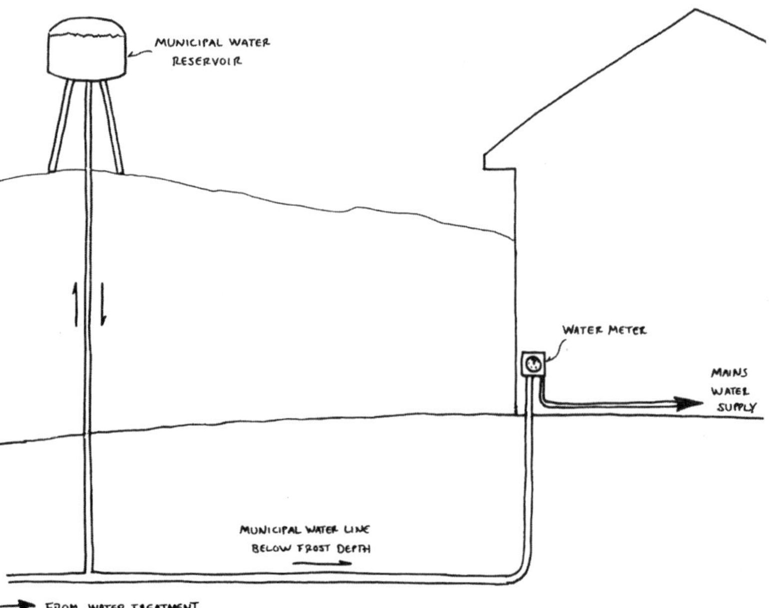

In combination, these issues make smart thinking about water systems a must for anybody interested in more sustainable building, as the dividends for lowering water use are double. The good news is that there is still a lot of "low hanging fruit" available, making significant water savings relatively easy.

Water systems are highly regulated in our building industry. Sources of water are subject to many levels of government jurisdiction that will regulate how, when, and where water is extracted and treated. All components of a drinking water system must meet the requirements of codes. In some regions, a drinking water system may need to be designed by a licensed professional, with each component specified and inspected.

What follows is an overview of available options when it comes to water systems. All are feasible, but not all are allowed by building codes. They may be combined in different ways to meet specific requirements, according to need, climate, personal preference, and local regulations.

Individual vs Municipal Supply

If your home is located in an area that is serviced by a municipal water system you may be obliged to connect to this water supply. Be sure to check local codes before pursuing any alternative to the municipal supply.

Surface water collection

System Components

- Foot valve and screened intake
- Tubing, buried to avoid freezing where applicable
- Land-based or submersible pump
- Pressure tank, if required
- Filtration as required

How the system works

Water is drawn directly from a lake, river, stream, or pond. This system is used for individual residences and entire municipalities. Water is drawn from the source as required, using the natural capacity of the water body as storage.

Water Quality Issues

Surface water is vulnerable to a wide range of natural and human contamination, and will almost certainly need extensive treatment in order to be potable.

Assessing water quality can be difficult, as the water will vary greatly in quality depending on season, weather events, human influence, and natural cycles and issues. Before choosing to use a surface water source, determine the origin of the water body and find out what kinds of activities happen upstream from your intake, in particular industries, sewage treatment facilities, and major highways. Each of these can introduce contaminants. If the water body is subject to seasonal flooding and/or drought, this will also affect quality by introducing new contaminants or concentrating those already existing.

Surface water should be tested regularly, as levels and types of contamination will change.

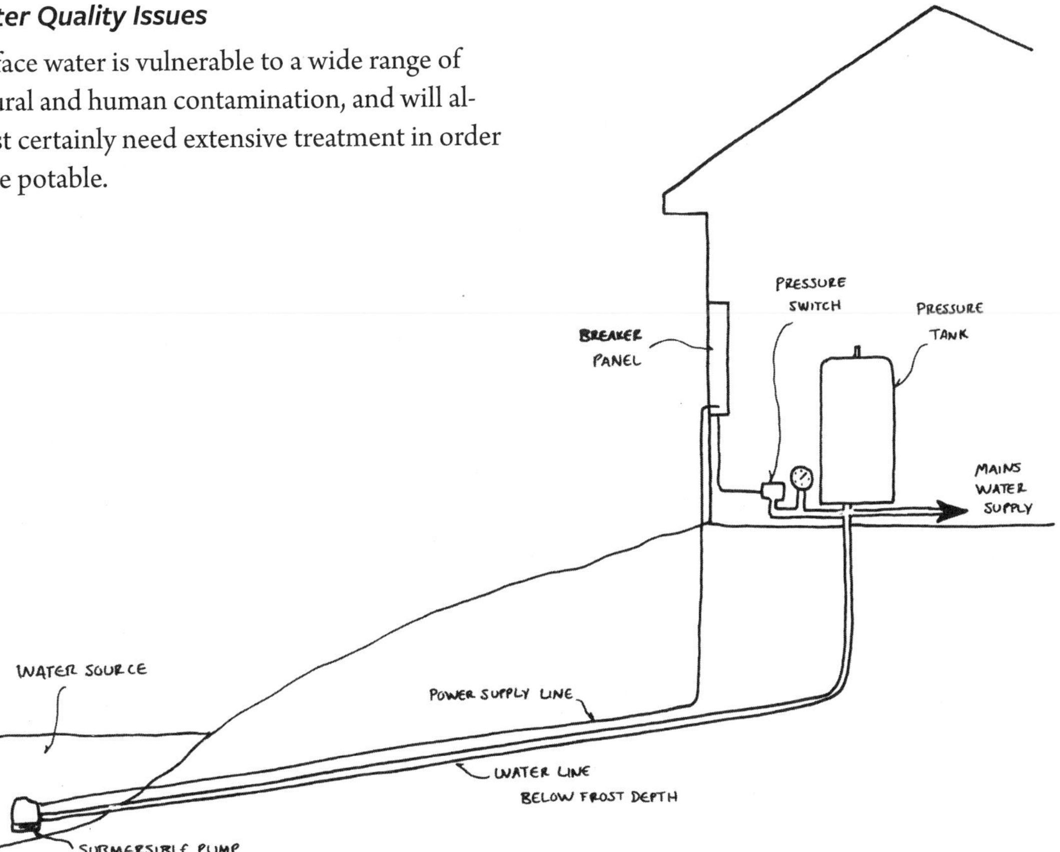

Well water collection

System Components

- Well casing (concrete for shallow wells, metal for deep wells)
- Submersible pump
- Tubing, buried to avoid freezing where needed
- Pressure tank
- Filtration as required

How the system works

There are two types of wells, but both operate on the same principle. An underground aquifer is accessed by digging or drilling an intake point into an area with sufficient flow to provide the required quantity of water.

- **Dug Wells** — These wells tend to be shallow, between 10–40 feet. A hole is excavated (by hand or mechanically), and the sides of the hole shored up and retained by sidewalls. Water in the ground is able to collect inside the well, and it is pumped or lifted from this reservoir.
- **Drilled Wells** — Used to access deep aquifers (between 40–400 feet), these wells are mechanically drilled (through any type of soil or rock) until sufficient water has been reached. A metal well pipe is fitted into the hole, and water from the aquifer fills some portion of this pipe. A submersible pump is lowered into the pipe and sits in the water.

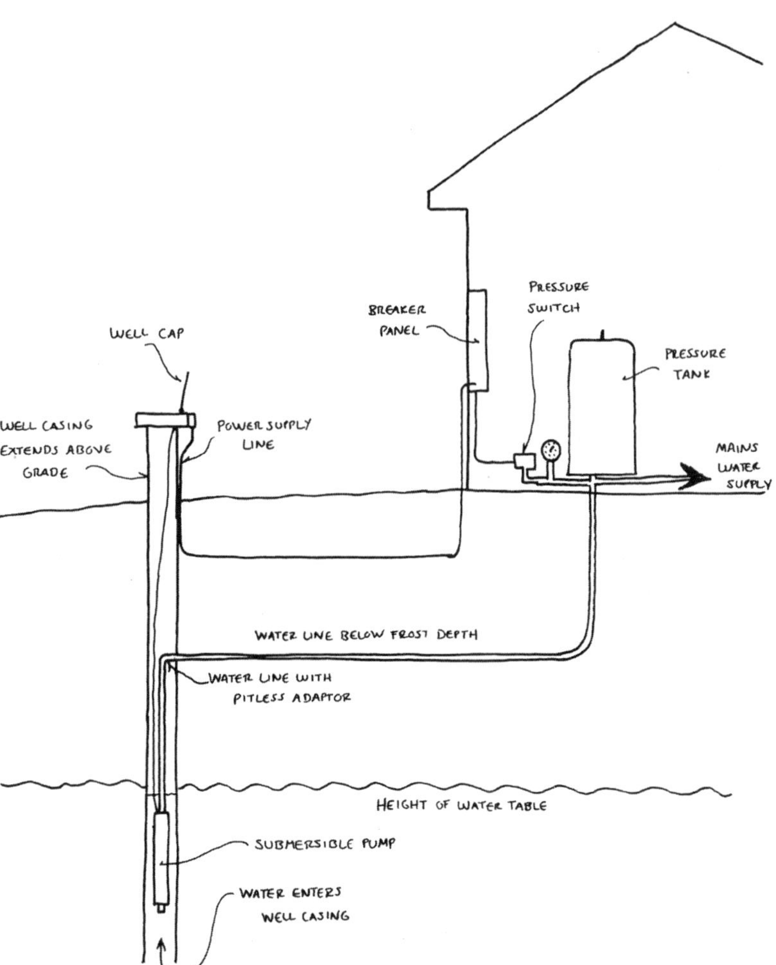

Water Quality Issues

Drilled wells may supply potable water with no need for treatment. Deeper wells tend to have fewer issues with bacterial contamination, but are often rich in mineral content. Depending on the composition of the rock around the aquifer, odor and taste of the water can be affected, and the interior of piping can suffer build-up of mineral scale. Treatment may be required to remove mineral content.

Drilled wells may be contaminated by sources far from the intake. The movement of underground aquifers is not well mapped, and water can travel long and circuitous paths. Deep wells tend not to be affected by seasonal changes, and a well that provides clean water is likely to continue to do so unless new human activity somehow affects it.

Rainwater collection

System Components

- Roofing of suitable material for collecting potable water
- Eavestroughs/gutters with screen covering
- Downspouts
- First-flush diverter
- Storage tank
- Land-based or submersible pump
- Filtration as required

How the system works

The roof area of a building is used to capture rainwater via gutters (eavestrough). Downspouts typically carry the water to a first-flush diverter, which directs a quantity of water from the beginning of a rain event away from the tank to prevent contaminants on the roof from entering the tank. The water is directed to a storage tank where it is held until required for use.

Water Quality Issues

Rainwater is typically very clean, unless contaminated by particularly heavy air pollution or toxic dust on the roof. Rainwater is distilled water and therefore has very low mineral content. Unless re-mineralized, it can have long-term health effects if it is the main source of drinking water. Low mineral content in the water can also cause leaching of mineral content from piping, which is of special concern if the piping contains fittings that may have lead and/or zinc content.

Re-mineralizing of rainwater can be achieved using a simple sand filter, or by partially filling the storage tank with sand and gravel.

The storage tank is susceptible to algae growth under the right temperature and light conditions. Dark, cold tanks are unlikely to support algae.

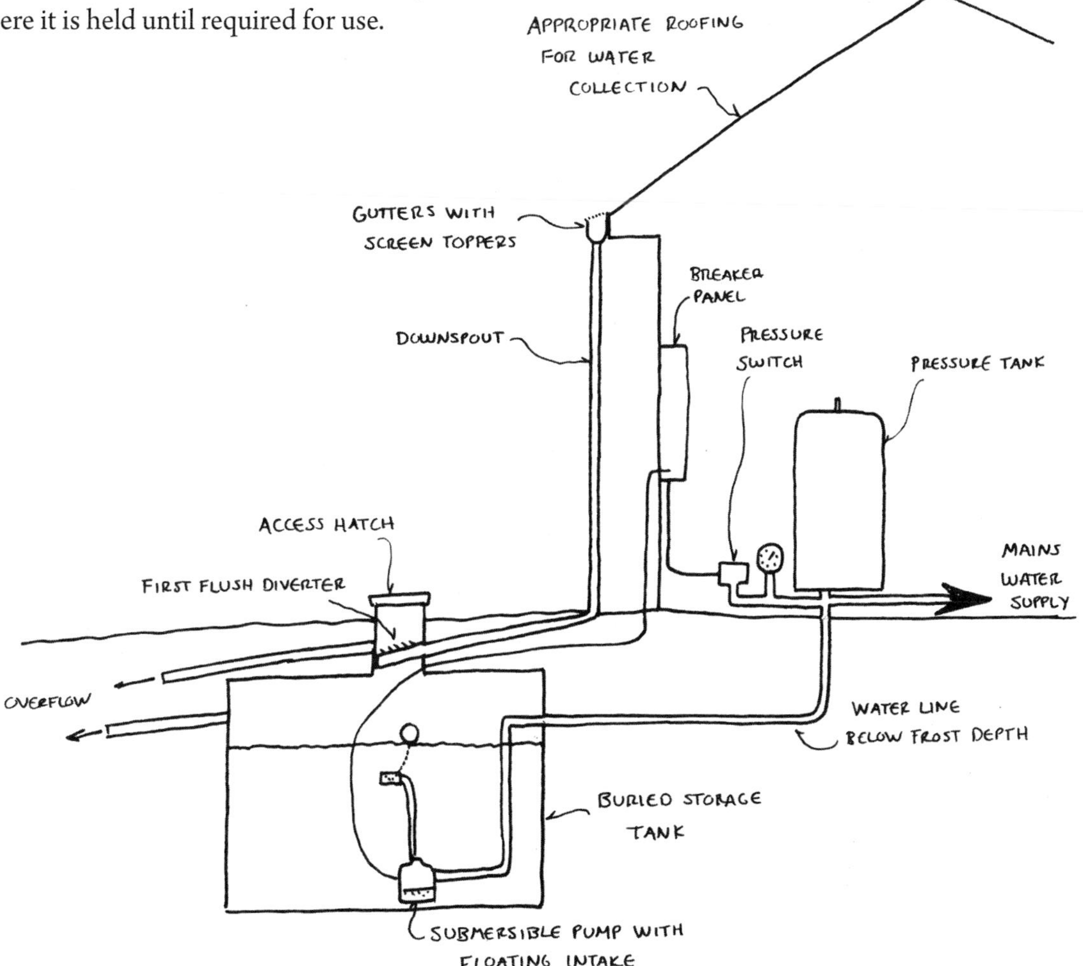

Desalinated water

How the system works

- **Distillation methods** — Include *vapor compression, multi-stage flash distillation,* and *multiple-effect distillation.* These methods use high amounts of heat to separate salt content from the water.
- **Membrane processes** — Include *reverse osmosis, electrodialysis reversal, nanofiltration,* and *membrane distillation.* These methods use pressure and a membrane capable of capturing salt while allowing water to pass through.
- **Solar desalination** — There are many variations of solar desalination, but most rely on evaporative processes using the sun's energy.

Desalination can be undertaken at a residential level, which would favor membrane or solar technologies, while municipal facilities use large-scale distillation or membrane plants.

There is a great deal of research and development going into improvements in existing technologies and exploring new ways of achieving desalination. Low-temperature thermal desalination and thermionic technologies are both promising. Many low-tech, homemade, or small-scale systems have been invented and are in use around the world at a residential scale. Small desalinators typically work by slowly processing water, which is then stored for use. A storage tank is therefore an integral part of the system.

Water Quality Issues

Desalination provides high-quality treated water.

Mechanical Systems: Wastewater

The development of modern wastewater systems has been based on only one objective: ensure that the user doesn't have to think about wastewater. It is an issue we collectively haven't wanted to consider. However, dealing properly with wastewater is a critical part of creating a sustainable building, and any home built to the status quo in this regard is contributing to environmental issues including water resource depletion, surface, groundwater and soil contamination, and high costs for municipalities and homeowners.

Our watersheds are contaminated, unfit for drinking and even swimming because of the mismanagement of our wastewater. Drinking water sources, crops, and soils can similarly be rendered toxic. Wastewater treatment typically represents 15–35% of the overall budget of municipal governments, and many municipalities are facing issues with aging sewer infrastructure that will require larger infusions of capital expenditures.

The very concept of "waste" water is at the root of this issue. The majority of what goes down our drains need not be considered waste. Certainly, the water itself should not be considered waste; it is one of our most valuable resources.

The regulatory framework for dealing with wastewater is the most restrictive and prescriptive aspect of most codes, and many homeowners are thus dissuaded from attempting to employ more sustainable strategies. The impetus for these regulations is understandable: a desire to minimize the harm caused by improper wastewater disposal and/or inadequate treatment. Unfortunately, many of the accepted solutions in current codes reinforce practices that are responsible for much of this harm, and discourage the use of alternative solutions that attempt to address these issues.

There are signs of positive change in this area. Rainwater catchment, greywater recycling, and composting toilets are beginning to find acceptance in some codes, and it is likely that the majority of codes will follow in the next decade or two.

Homeowners wishing to pursue sustainable wastewater strategies should familiarize themselves with local regulations and if proposing an alternative solution be prepared to absorb extra time and possibly extra cost in the planning and construction process.

What follows is an overview of available options when it comes to wastewater systems. All are feasible, but not all are allowed by building codes. They may be combined in different ways to meet specific needs, according to need, climate, personal preference, and local regulations.

Municipal wastewater treatment

How the system works

A building's drain system empties into a publicly maintained sewer system, a network of subterranean pipes that flow (or may be pumped) to a centralized wastewater treatment facility. These facilities are typically located close to a natural water body as a natural low-elevation point and as a place to discharge treated water.

The operations and processes at wastewater treatment facilities vary depending on local practices, regulations, and environmental conditions, but follow a similar procedure:

- **Primary treatment** — Wastewater is collected in holding tanks or settling ponds so that "scum" (grease, oil, soaps) can rise to the surface and solids settle to the bottom. Scum is removed for separate treatment or landfill disposal. Solids (or "sludge") are typically sent to a digester, where anaerobic activity helps to break down dangerous pathogens.
- **Secondary treatment** — En route from primary to secondary treatment, water is aerated. Secondary treatment processes vary widely between facilities. In some cases, additional settling and mechanical filtration are the only remaining step before discharge to waterways. Sometimes the water is given adequate conditions to allow microorganisms to treat dissolved and suspended biological matter.
- **Tertiary treatment** — There are a variety of possible tertiary treatments, including biological processes, chemical treatment, and microfiltration. After tertiary treatment, water is released to the environment, or may be directed toward other uses such as irrigation or industry.
- **Sludge treatment** — In every cubic meter of wastewater, there are between 80–220 grams (about 3 to 8 ounces) of solids.[18] This aspect of treatment facilities is often overlooked when volumes of treated water are discussed. Sludge is subjected to different treatment processes, from landfill disposal to digestion, drying, and finishing. Sludge is often highly contaminated and only partially treated before being used as fertilizer, resulting in contaminated surface water and soil.

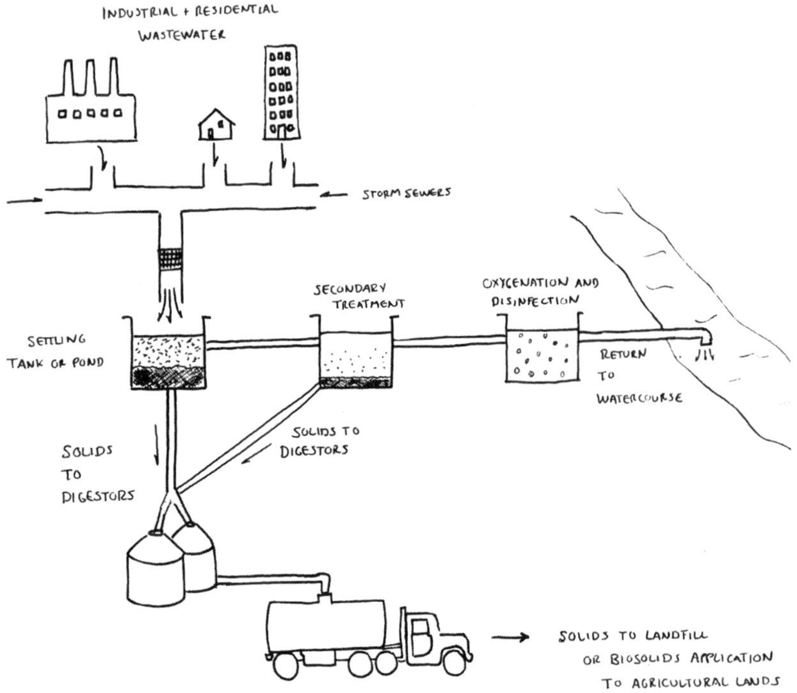

Any discussion of municipal wastewater treatment must include a reminder that many municipalities provide little or no treatment of wastewater before it is ejected into waterways. According to Environment Canada in 2009, 18% of the population has only primary treatment or less.[19] The Clean Water Act (1972) and supporting grants in the United States eliminated raw sewage discharge by the 1980s, but, according to the EPA, 10 million people are still served by systems that provide only primary treatment. In both countries, the number of people served by systems that include some form of tertiary treatment prior to discharge of wastewater is remarkably small. If you have high environmental goals for your project, research the treatment process of your municipal wastewater treatment system.

Septic systems

How the system works

Sewage discharge from the home is carried to a septic tank equipped with one or two chambers. The inlet to the tank is equipped with a baffle that forces incoming solids to the bottom of the tank. Further baffles are provided to keep floating scum from clogging the exit pipe. After initially filling, every incoming quantity of sewage forces an equal amount of effluent out the exit pipe and into a series of perforated pipes known as the weeping bed. Here the effluent is discharged into the ground where it percolates in the soil and is remediated to whatever degree the biological conditions in the soil can provide.

Tanks and weeping beds are sized to meet standard flow rates for the number of bedrooms in the home. Soil tests are often required of the weeping bed area to ensure adequate percolation rates.

A degree of anaerobic secondary treatment happens within the tank, but the majority of treatment occurs in the biolayer of the soil in the weeping bed. Very little breakdown of solids happens in the tank, and the eventual accumulation requires removal. These solids are often taken to municipal wastewater treatment facilities (see "Municipal wastewater treatment") or to landfill.

Tertiary treatment — Many alternative wastewater treatment systems — including constructed wetland, peat beds, aerobic tanks, soil air injection, and bio-filters among others — are extensions of the basic septic system, using the same type of tank and weeping bed setup, but introduce some form of tertiary treatment between the tank and the bed, helping to reduce harmful bacteria counts in order to deposit cleaner water into the soil.

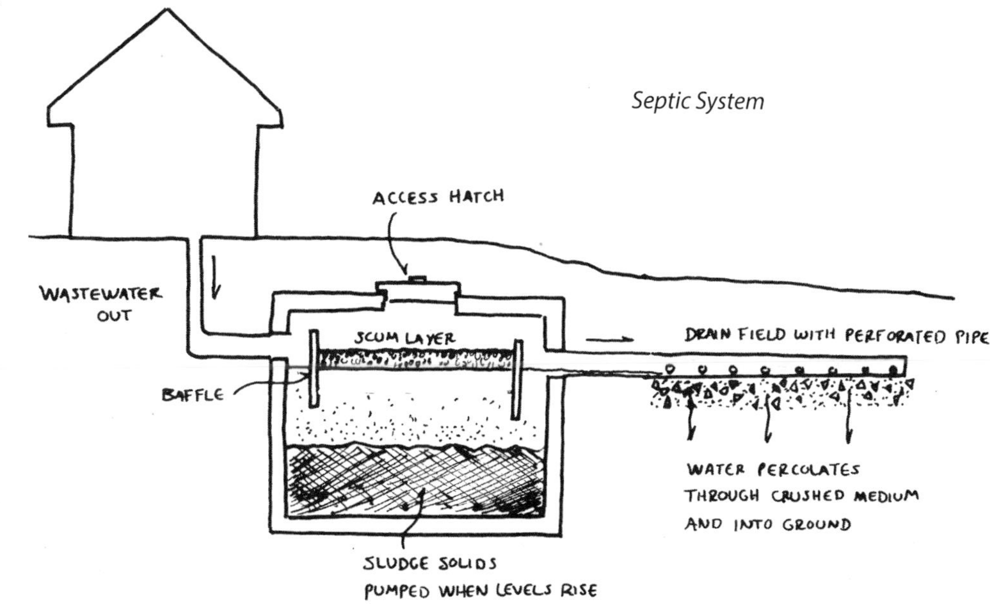

Septic System

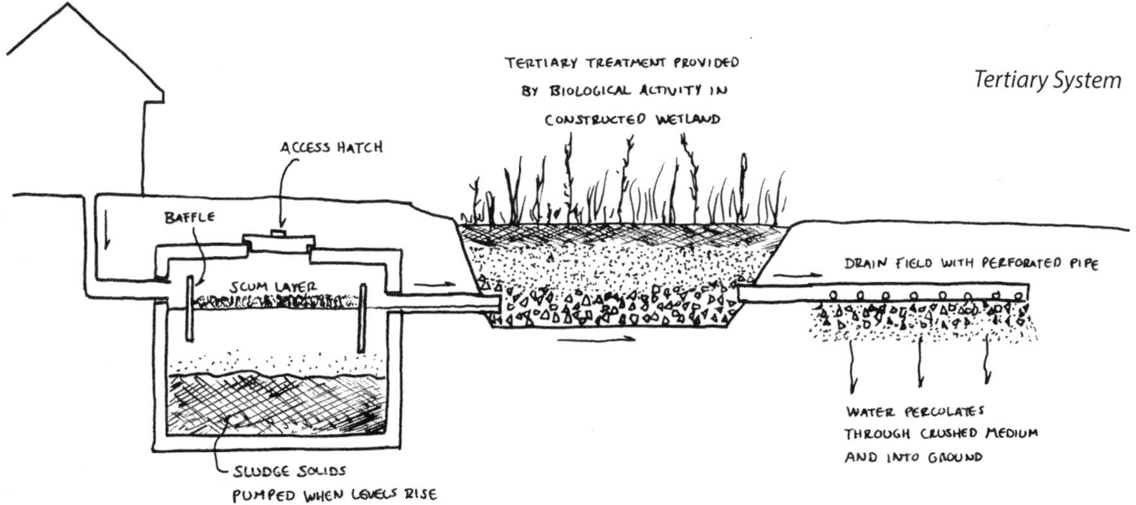

Tertiary System

Composting toilets

How the system works

Composting toilets collect urine and feces — referred to as *humanure* in the rest of this chapter — and provide complete treatment of this material on site until it is transformed into useful compost or humus.

There are three common types of composting toilet:

Bucket toilet — This low-tech version of the composting toilet uses a bucket or similar portable receptacle placed under a seat/container to receive humanure deposits. Sawdust, wood shavings, chopped straw, or other form of cellulose material is used to cover each deposit in the toilet, helping to reduce odor, absorb urine, and provide aeration. Once full, the bucket is emptied into an outdoor compost heap. Here the material is layered and mixed and covered with more cellulose material, providing the right conditions for the natural conversion to compost/humus.

The indoor toilet construction is usually provided with passive or active ventilation, but no water connection or flushing action is used.

Self-contained toilet — These units provide a seat over an integral composting tray in a single, self-contained structure. Humanure deposits are received in the tray and provided with the appropriate conditions for composting action within the unit. They all use some form of mechanical ventilation to reduce odor. Excess urine may require a separate handling system, or heat may be used to speed evaporation. Due to limited storage capacity, these toilets most often use some form of mechanical action and/or acceleration for the composting process, and are only suitable for low numbers of users or for intermittent use.

The compost tray is removed from the unit when processing is complete or when the tray is full. It is often necessary to have an outdoor compost heap to receive material from these units, as it can prove difficult to complete the composting process within the unit.

Some models of self-contained toilet use chemicals or high heat to "cook" the humanure into a benign state. The material from these toilets is not useful compost, as the biological activity that creates rich, useful soil has been killed off.

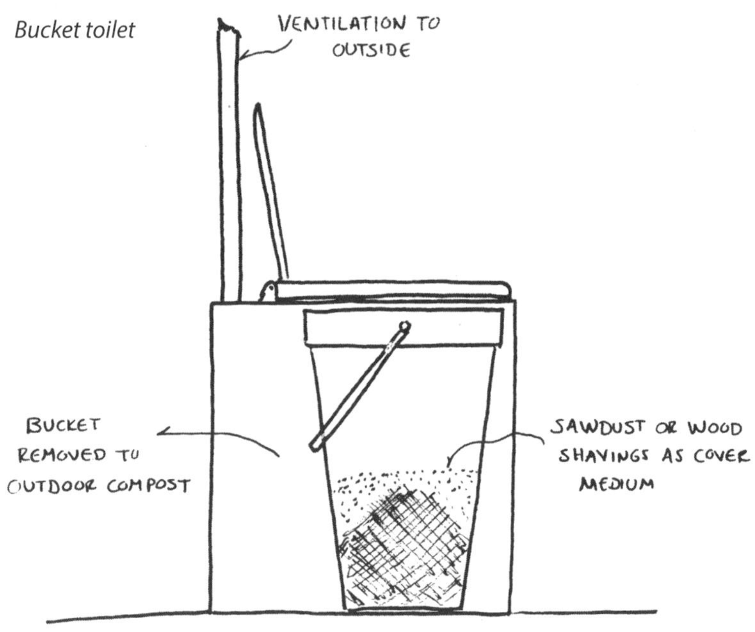

Bucket toilet

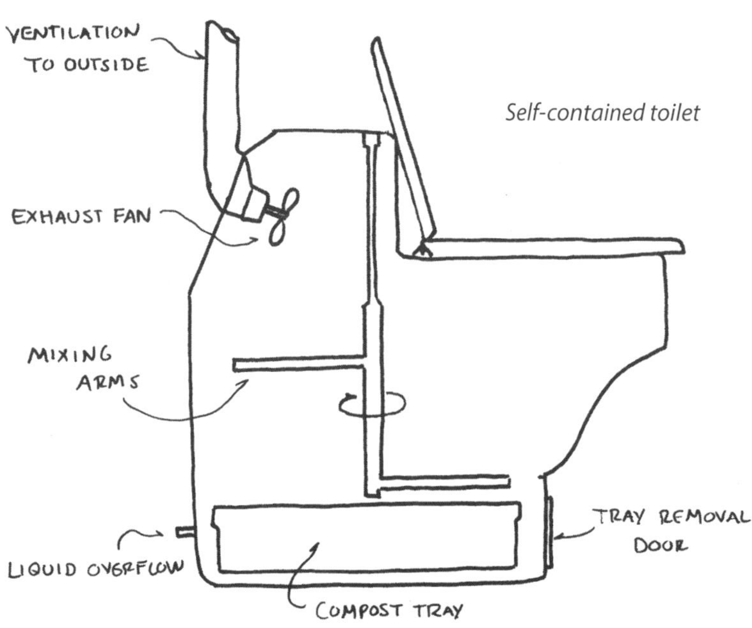

Self-contained toilet

Remote chamber toilet — A toilet (dry chute, low-water, or vacuum flush) directs humanure to a large, enclosed chamber. The chamber is of sufficient capacity and design to contain and process a high volume of compost.

Units vary in how humanure is handled. Some use heat and/or evaporation to rid the chamber of excess urine and water and speed the composting process, while some retain and process all material. Mixing or stirring capabilities, misting sprayers, and rotating trays are options offered by certain manufacturers.

Some units gather excess urine after it has passed through the bulk material in the chamber and retain this liquid as a high-quality fertilizer. This makes best use of the potential value of all material entering the toilet, as up to 80% of the nutrient value in toilet waste is in urine. Once transformed into nitrites and nitrates after passing through the biologically active compost solids, the fertilizer can be a safe and low-odor fertilizer.

All chamber style toilets provide humanure with enough time and adequate conditions to allow the composting process to fully convert to compost before being removed from the unit. These are the only units that do not require additional outdoor composting capacity.

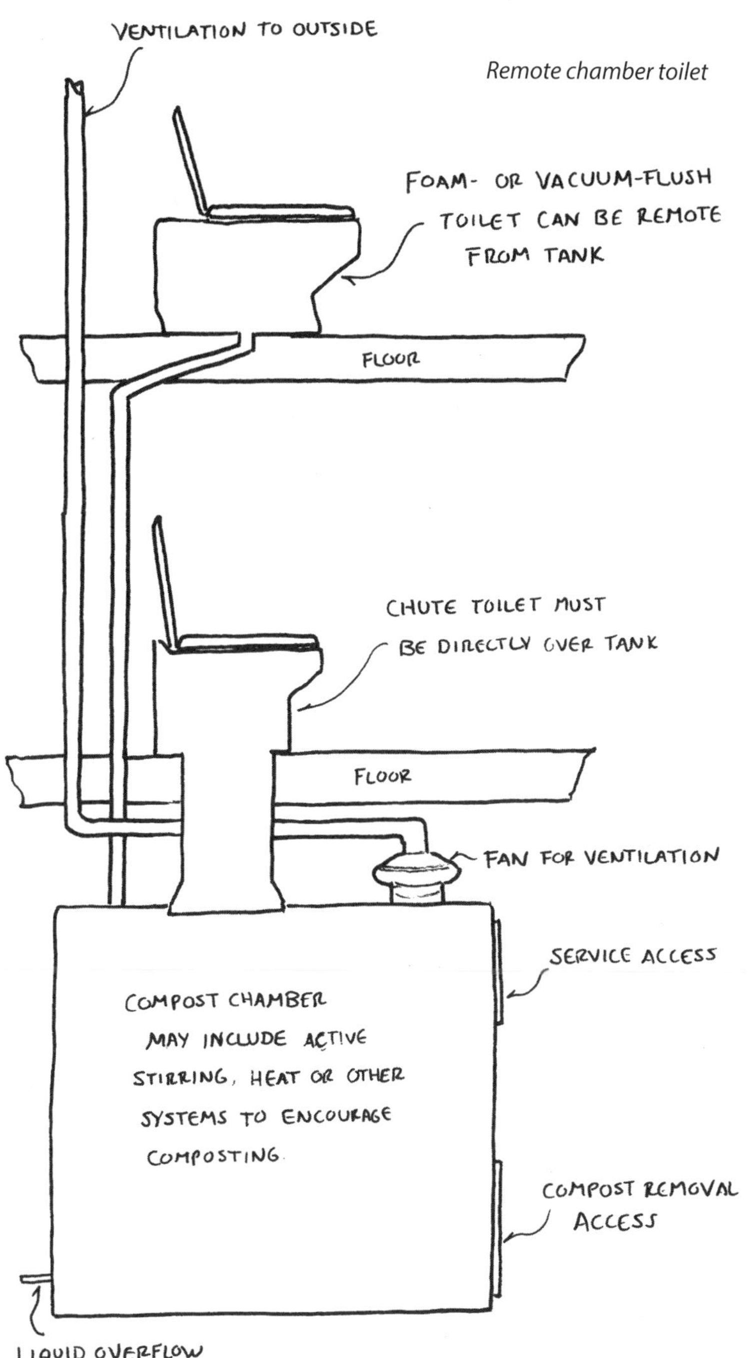

Remote chamber toilet

Greywater systems

The use of composting toilets results in the separation of "black water" (containing human waste) from all other wastewater, which is referred to as *greywater.* In residential settings, this water is typically quite "clean" and may be able to be treated in a few different ways:

Weeping fields — Greywater is disposed directly into the ground using a process similar to septic systems, in which a holding tank allows for settling of solids and rising of scum, and perforated drainage tiles allow the water to be released into the ground. Microorganisms in the soil deal with any harmful bacteria. Some systems include a biological filter in place of a septic tank to help reduce costs.

Outdoor irrigation — Greywater is disposed into a weeping field that provides subsoil irrigation for outdoor trees and/or gardens. Systems have the same components as weeping fields, with the drainage tiles intentionally located to provide water to plants.

Indoor irrigation — Greywater is directed into planter boxes that use stone aggregate to allow water to flow sub-surface beneath growing medium and plants. Microorganism colonies in the aggregate, soil, and plant roots help to clean the water. Some systems retain all the greywater within the system, providing adequate storage capacity to allow transpiration through the plants to release water to the atmosphere. Other systems direct greywater into an outdoor weeping field after it has flowed through the planter boxes.

Greywater recycling — Greywater is directed through filters into a storage tank, where it is reused for toilet flushing, irrigation, and other uses in the home. The type of end use for recycled greywater is dependent on the degree of filtration provided. Ozone or chlorine treatment is often required for greywater recycled for indoor use. Overflow from these systems is disposed through weeping field or irrigation systems.

Outdoor irrigation

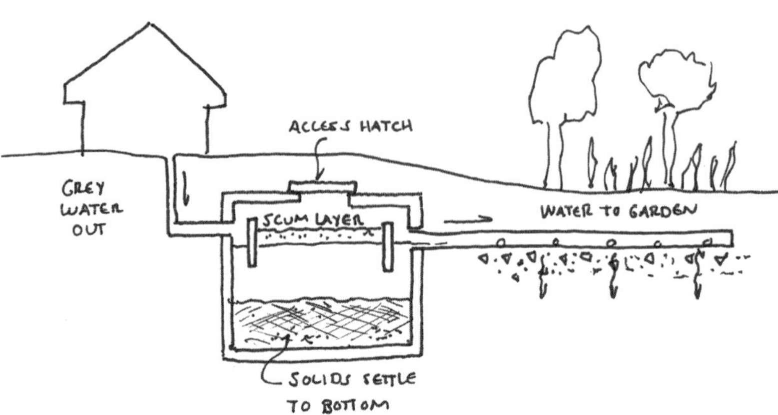

Indoor irrigation

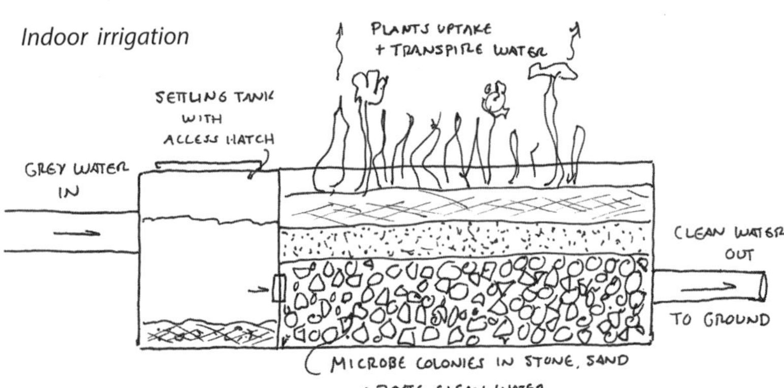

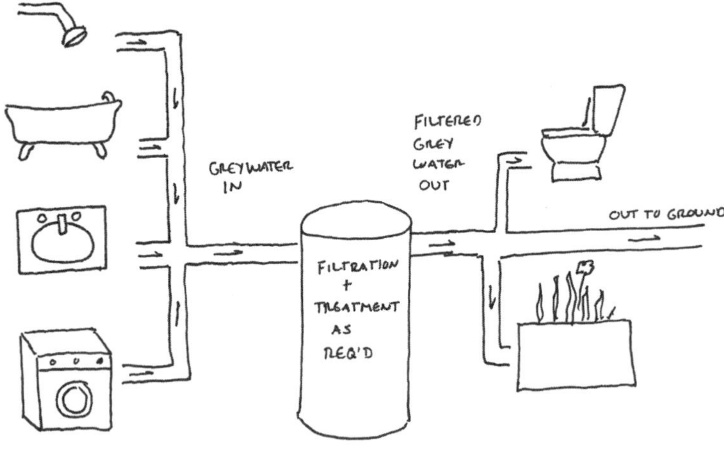

Greywater recycling

Mechanical Systems: Electrical Generation

In the short space of a century, electrical energy has gone from a rare novelty to an absolute necessity. For lighting, heating, refrigeration, communications, or work-reducing appliances, electrical energy is ubiquitous in our homes.

Homeowners wishing to make use of electrical energy have three major sourcing options to meet their power needs.

Grid Power — Most homeowners receive their energy from public or private utility companies, whose large, centralized generating stations produce high volumes of electrical energy that are distributed through a network of transmission lines that has come to be known as "the grid."

There are many kinds of electrical generation on the grid, including:

- **Hydroelectric** — The power of falling water is used to spin turbines that generate electricity.
- **Fossil fuel plants** — Heat from the burning of fossil fuels (including coal and natural gas) is used to create steam that is used to spin turbines that generate electricity.
- **Nuclear plants** — Heat from the fission of atoms is used to create steam that is used to spin turbines that generate electricity.
- **Wind turbines** — The power of wind is used to spin turbines that generate electricity.
- **Solar thermal** — The heat of the sun is concentrated and used to create steam that is used to spin turbines that generate electricity.
- **Photovoltaic** — Photons from the sun are used to displace free electrons on a silicon wafer to generate electrical current.

Regionally, different balances of these sources make up the overall available energy. It can be difficult for a homeowner to know the exact source of the electrical energy being used in the home without research.

The majority of electrical power on the grid is generated by means that have major environmental impacts that are well documented. Centralized production is also extremely inefficient, with overall losses in production ranging from 30–70% and transmission losses around 8%, which means that only a fraction of the energy value of the fuel being used is actually making it to the end user.

A small number of private utility companies offer homeowners a means of purchasing renewable energy from the grid. Under these programs, the amount of electricity used by a homeowner is put onto the grid by the private utility, offsetting the amount of "dirty" power needing to be generated.

Grid-tied — The grid is a two-way power highway; electrons will move from a point of generation to a point of consumption in either direction along a wire. Until recently, utility companies treated the grid like a one-way street running from central production facilities to end users. In recent years, the value of "distributed generation" has become evident, and many utilities have begun to allow for small-scale production by homeowners and businesses, often under direction from governments.

"Grid-tied" homeowners and businesses produce power (typically with renewable sources like photovoltaic, wind, and small-scale hydro) under a variety of different contracts and terms. Generated power is put onto the grid, and the owner of the generator receives money or credit for that power. If the owner also requires electrical power, it is supplied from the grid. In this way, a homeowner can become a "net-zero" electricity user, generating as much power as is used.

These grid-tied systems offer numerous benefits to both homeowners and utility companies. Homeowners can provide a portion of their electrical needs (from a small percentage to overproduction for profit), yet not be reliant on having to store and manage their own power supply, as would be necessary for off-grid systems. Renewable sources like solar and wind produce a lot of power sometimes, and no power at other times. With the grid as a "buffer" the home is never without power when needed. Utility companies benefit from having homeowners and businesses put out the investment for new production that can collectively offset the need for expensive new large-scale generating capacity, and increase the percentage of clean renewable energy in their mix.

Distributed generation helps to reduce the high production and transmission losses associated with centralized power plants by reducing the distance from point of production to point of use, and creates a more resilient grid less susceptible to massive outages when a large power plant goes offline.

Off-Grid — A much smaller number of homeowners generate and manage their own electrical energy, functioning independently in off-grid homes. These systems typically rely on a bank of batteries to provide chemical storage of electrical energy, which can be charged by the home energy system as power is available and drained when power is required in the home.

Off-grid systems reward energy-efficient home design and conservative power use within the home. If energy demands are low, this type of system can be affordable and reliable. As demand rises, the systems grow in size, complexity, and cost.

The addition of battery storage requires a space for housing the batteries and associated controls, and the owner must maintain the batteries and ensure the balance of power in and out of the system. New battery technologies are now reaching the market, and developments in energy storage may make off-grid systems — or grid-tied systems that include in-house storage — more affordable and practical.

Off-grid generation systems typically have higher capacities than grid-tied systems to accommodate for the variable nature of renewable energy; daily and seasonal swings in generation can require overproduction and large storage capacity to ensure power is available through times of low or no production. Off-grid systems provide a high degree of independence and resilience, with little or no reliance on outside sources of production or delivery.

Photovoltaic power

Energy Source

- Solar radiation

Ch:3

System Components

- Photovoltaic panels
- Mounting racks
- DC to AC inverter
- Connectors, junctions, and balance of system components
- Storage batteries (if required)

How the system works

The *photovoltaic effect* describes the ability of photons of light (predominantly from the sun) to excite electrons into a higher state of energy, allowing them to act as charge carriers for an electric current.

Most photovoltaic systems use flat cells of silicon that have been "doped" with an additional electron on one side and one less electron on the other. Photons strike and excite the extra electron, causing it to "jump" to conductors embedded in the cell, where it completes the circuit wired to the cell. Multiple cells are linked together as modules (panels), which are protected by a glass covering and typically installed in an aluminum frame. Multiple panels can be wired together to provide a desired output.

PV modules produce *direct current* (DC). In most systems, an inverter must be employed to change DC to *alternating current* (AC), which is the type of current used for grid power and household appliances.

System Output

Modules from different manufacturers will have varying output depending on cell size, number of cells in the modules, and efficiency of the cells. Modules are rated for their output in watts in peak sun conditions. Typical output for current photovoltaic technology is in the range of 1000kWh output per year per 1kW peak module sizing, or an average continuous output of 114W.

Current PV technology converts between 13–21% of the available solar energy to electric energy.

Wind turbines

Energy Source

- Wind movement

System Components

- Airfoil blades
- Generator
- Directional tail
- Tower base
- Balance of system as required
- Storage batteries, if required

How the system works

Airfoil blades are attached to the shaft of a generator. The force of moving air causes the blades to spin and the generator to produce current. The blades and generator are mounted on a pivot at the top of a tower that places the unit in clean airflow, undisturbed by obstructions that cause turbulence in the wind.

Most residential units produce "wild" power, with output levels that vary based on wind speed. Controllers are used to even out the power delivery. Power may be produced in either AC or DC current, depending on the design of the unit. Some form of wind speed limiter will be used for every turbine, as extreme wind conditions can cause severe damage to the unit.

System Output

Wind generators are rated according to their output at a particular wind speed. The wattage advertised is not a static output, and manufacturers do not all use the same wind speed for their ratings. This makes direct comparisons difficult.

More important than any advertised rating is the power curve for the unit, which shows how much power is produced at any given wind speed. Choose a unit that has the best possible output at the average wind speed at the installation site.

Micro-hydro turbines

Energy Source

- Flowing water

System Components

- Turbine (many styles to suit different conditions)
- Generator
- Penstock (water intake)
- Tailrace (water release)
- Valves and controls as required

How the system works

A turbine uses the flow and pressure of falling water to turn a generator to produce electrical current. A waterway must have a suitable amount of *head* (elevation between inlet and outlet points) and *flow* (quantity of water) to make a useful amount of power. A *penstock* diverts some water from the high point in the system and directs it to the turbine. The pressurized water is passed through the turbine causing it to spin and turn the generator to produce current. The water leaves the turbine and rejoins the flow of the river or creek.

There are numerous styles of turbine, each one suited to particular head and flow characteristics. Valves are installed to shut off flow for servicing.

System Output

Output figures are based on available head and flow and the efficiencies of the piping and turbine. Very low head, low flow systems can generate as little as 20 watts, while the upper limit of micro hydro is generally accepted to be 100 kilowatts. Larger output systems are considered to be full-scale hydro electric projects.

A water turbine system produces power constantly, unlike solar and wind systems. Even relatively low output can add up to a significant amount of power.

Mechanical Systems: HVAC

Modern heating and cooling systems are often complex; high-performance devices give us fingertip control over indoor climate that would have been unthinkable less than a century ago.

Though most heating and cooling devices are intricate systems, it is quite easy to understand the basic technology behind each of them. It is worthwhile as a homeowner to understand these systems, and not leave it to company reps or installers to provide selling points.

HVAC choices will be dependent on all your design and material decisions, as these will determine how much heating and cooling energy is required. The overall efficiency of your home will make different system options more or less feasible. The selection of heating and cooling systems should be done in conjunction with the energy-modeling phase of the project.

Passive solar heating

Heat Production and Delivery

- Solar heat production
- Passive heat transfer
- Passive air delivery

System Components

- South-facing glazing
- Thermal mass, as required

How the system works

Passive solar heating is not technically a heating system, rather it is a design approach that has space-heating benefits. Windows are arranged on the south-facing side of the building such that they are 20–30% of the floor area of the room behind. They are positioned to receive the full penetration of the low winter sun, and provided with static or active shading to prevent penetration of the high summer sun. Solar radiation enters the windows and warms the air and the mass in the room. Good passive solar design can also exclude direct warming from the sun in the summer months, reducing or eliminating the need for a cooling system.

System Output

Output can vary greatly depending on a number of factors:

- Available solar radiation
- Latitude
- Transmission rating of glazing
- Climatic conditions (hours of clear sunlight)

In northern climate studies, effective passive solar design has been shown to provide a solar heating fraction of 10–35%. While this is far from the full heating load, it is entirely free and a significant reduction in the requirement for other heating systems.

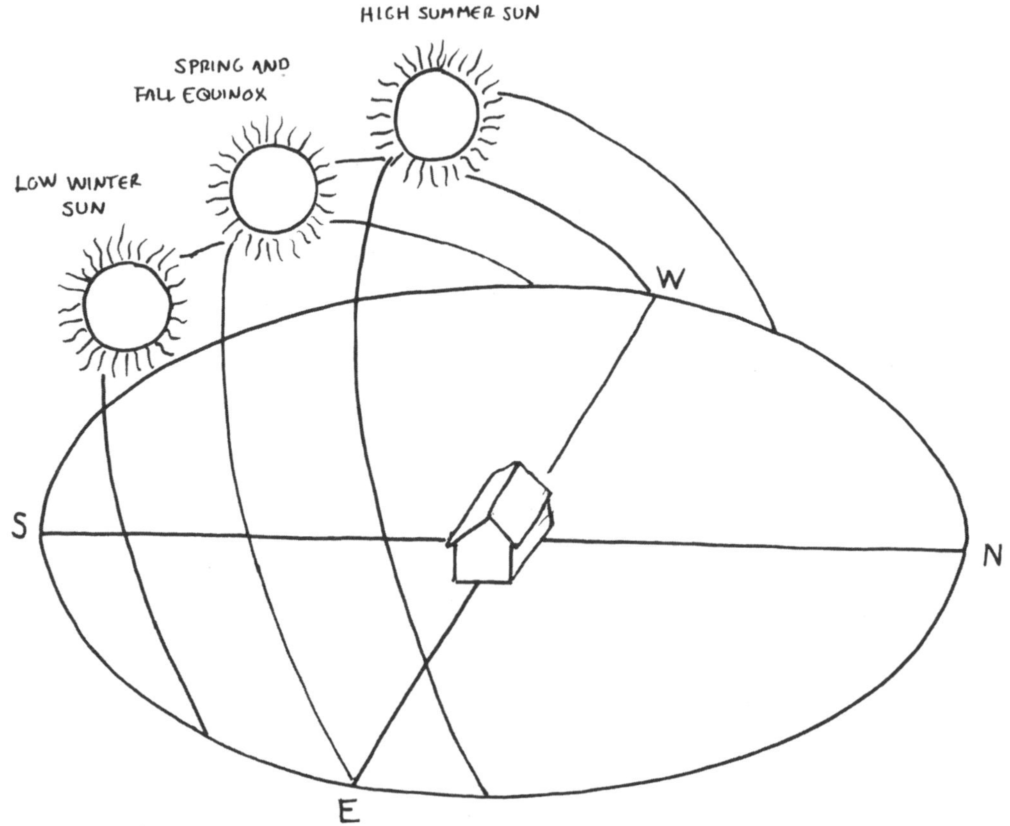

Solar hydronic heating

Heat Production and Delivery

- Solar heat production
- Active or passive heat transfer
- Hydronic or forced air delivery
- Can be used for domestic hot water heating

System Components

- Solar thermal collector (flat plate or evacuated tube)
- Circulation pump and controls
- Storage vessel
- Heat exchanger for delivery system

How the system works

Solar heat is absorbed by the large surface area of the collector plate or fin, and is transferred to a hydronic fluid. This fluid is pumped (or sometimes moved by convection) to a storage vessel where it deposits its energy via a heat exchanger to a tank of water. A thermostatically controlled pump circulates the hot fluid in the tank directly to the home's heat delivery system or through a second heat exchanger.

The heated water is stored in an insulated tank or vessel, allowing for use of the heat when required, not only when the sun is shining. Tank capacity and insulation levels must be designed to provide an adequate storage-to-production ratio so that heat is available when required.

A careful design process must be used to determine how much heat can be reliably delivered via a solar hydronic system. This will be the "solar fraction," which can range from 100% in some climates to as low as 25% in others. The cost-effectiveness of the system must be compared to the solar fraction and the cost of backup or complimentary heating.

Solar hydronic heating systems will overproduce heat during the summer months. If there is no adequate "dump" for this heat, the system will need to be drained and/or shuttered in the warmer months.

System Output

A typical solar thermal collector is rated at 10,000–50,000 Btu per square foot of collector per day. The lower figure corresponds to cloudy days and the higher figure to sunny days. Actual outputs are dependent on collector type, solar angle, and exposure. Site-specific estimates can be prepared to help with system design.

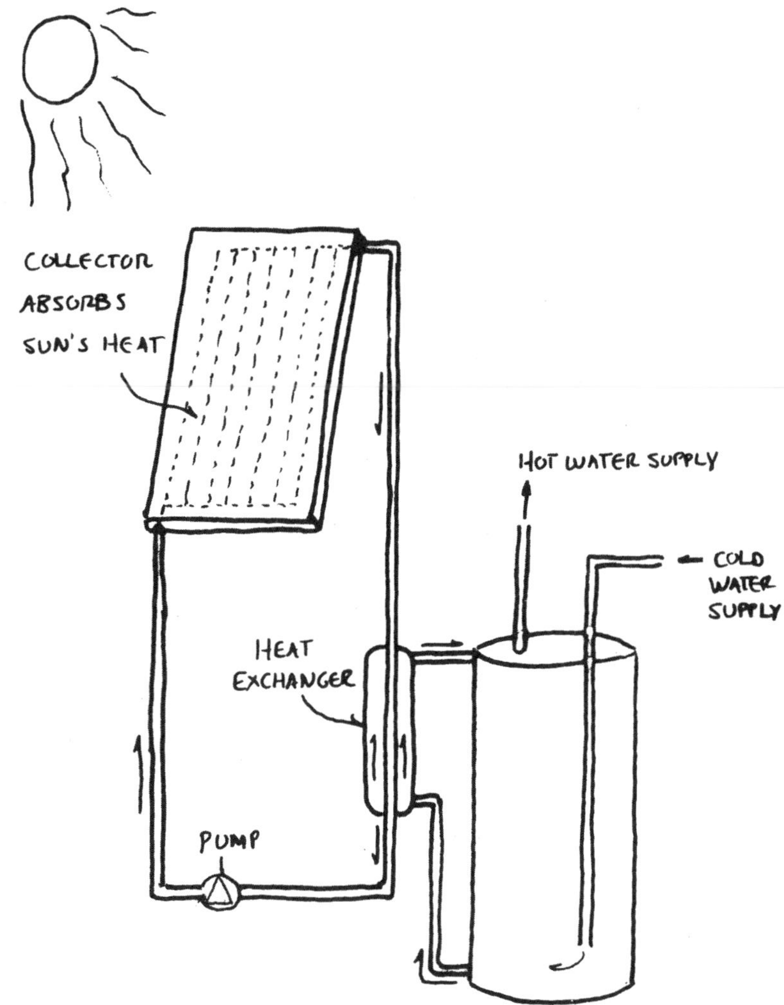

Solar air heating

Heat Production and Delivery

- Solar heat production
- Active or passive heat transfer
- Passive or forced air delivery

System Components

- Solar thermal collector (glazed or unglazed)
- Ductwork
- Controls and fans, as required

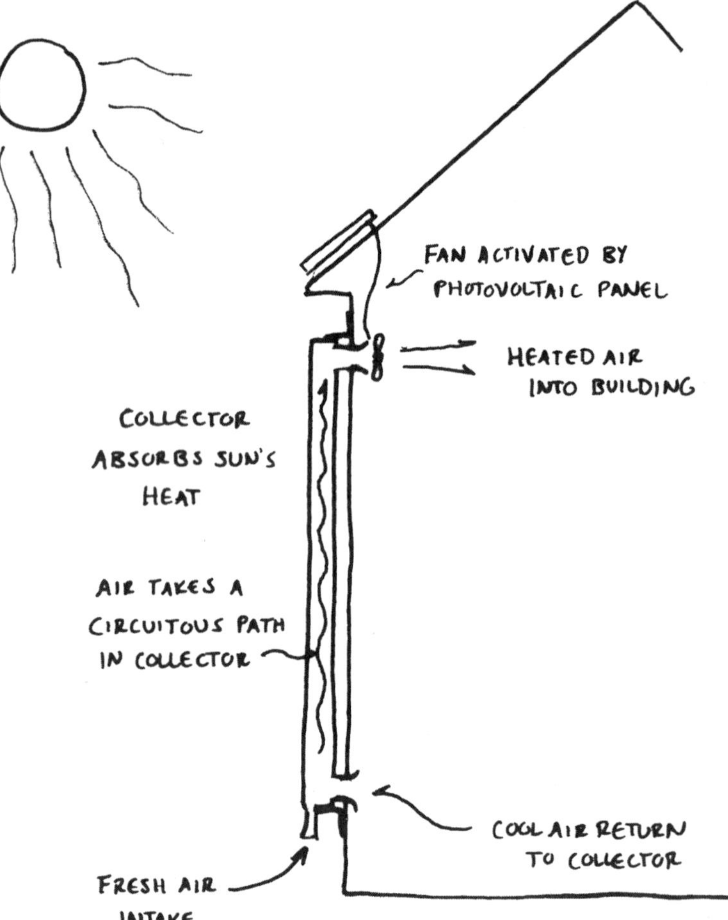

How the system works

Solar radiation is collected in glazed or unglazed collectors, in which air is heated in a plenum (chamber for heating air), often employing a circuitous or perforated pathway, and ducted into the building. Some systems use simple convection principles, and others use a fan and ductwork to move air to a desired location in the building.

The air being heated in the panels can be fresh air from outdoors, or it can be recirculated indoor air. It is possible to create a system where the air source can be selected on demand. Heated outdoor air is likely to enter the building at lower temperatures than recirculated air, but can provide much-needed fresh air in a tightly closed home in the winter without adding much strain to the heating system.

Most systems are designed to be inexpensive and simple. Solar fraction may be relatively low, but so are cost and complication.

System Output

There is a wide range of solar hot air collector designs, and output rates are highly variable.

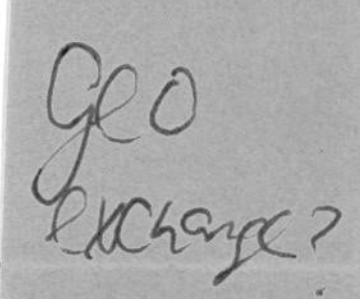

Ground source heat pumps (GSHP)

Heat Production and Delivery

- Heat pump production
- Air or hydronic delivery
- Can be used for domestic hot water heating

System Components

- Ground loops (buried horizontally in deep trenches or vertically in drilled wells) and circulation pump
- Heat pump unit with compressor
- Air or hydronic delivery system
- Controls as required

How the system works

There are two basic arrangements for GSHPs:

- **Horizontal Ground Loop** — The collection loop for the heat pump is placed in trenches that are dug to a depth that is below the frost line. Horizontal loops can also be submerged in bodies of water below the expected ice depth.
- **Vertical Ground Loop** — The collection loop for the heat pump is placed in one or more vertically drilled wells that are capped and grouted to protect groundwater.

The ground loops collect or disperse heat (depending on whether the GSHP is in heating or cooling mode) in the ground. The length of the tubes is designed to ensure that the fluid in the pipes has enough contact time with the ground to get input temperatures that are constant. The more Btus required of the system, the longer the ground loops. Ground loops can be closed loops that circulate a heat-exchange fluid, direct exchange loops in which refrigerant from the heat pump is sent through the ground loop or, more rarely, open loops where groundwater is drawn into the loop and then discharged in a different well or location in a body of surface water.

Fluid from the ground loops imparts its temperature to the refrigerant on one side of the heat pump, and the compressor puts the refrigerant through the heat pump cycle to either create or extract heat for the home.

The output side of the heat pump has a heat exchanger that transfers temperature to an air or hydronic delivery system, as required by the home's thermostat.

System Output

Independent testing of ground source heat pump units by Natural Resources Canada showed output for a wide array of residential units ranged from 8.7 to 12.8 Btu/hr/watt, or a *coefficient of performance* (COP) of 2.6 to 3.8.[20]

It is important when considering COP figures for heat pumps that the electrical energy required to run the circulating pump is often excluded from these figures, making them look more positive than the reality in use.

In North America, GSHPs are often rated by "tons" of output. A ton is 12,000 Btu/hr, and residential units typically range from 0.75–5 tons of output.

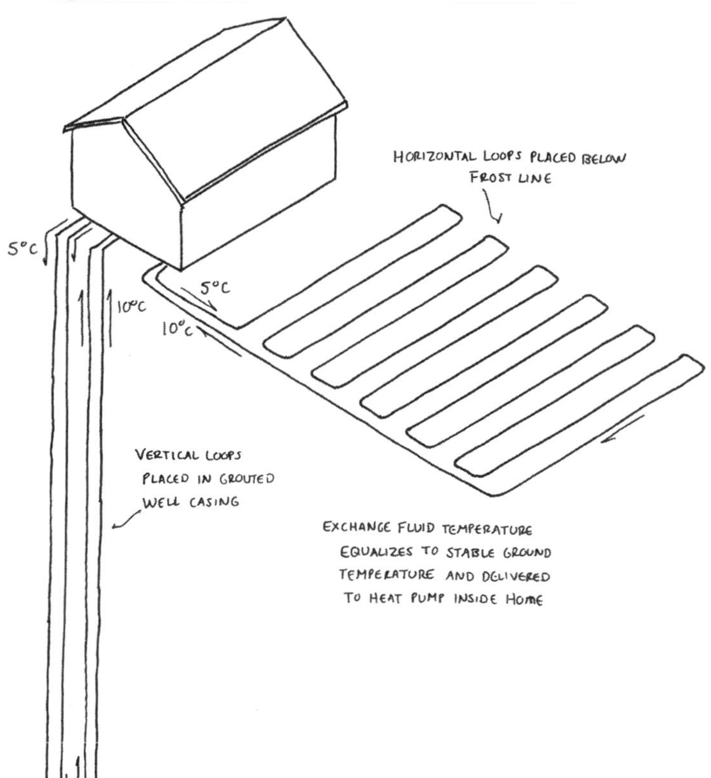

Air source heat pumps (ASHP)

Heat Production and Delivery

- Heat pump production

System Components

- Heat pump (placed outside the home)
- Heat exchanger/plenum (inside the home)
- Air or hydronic delivery system
- Controls as required

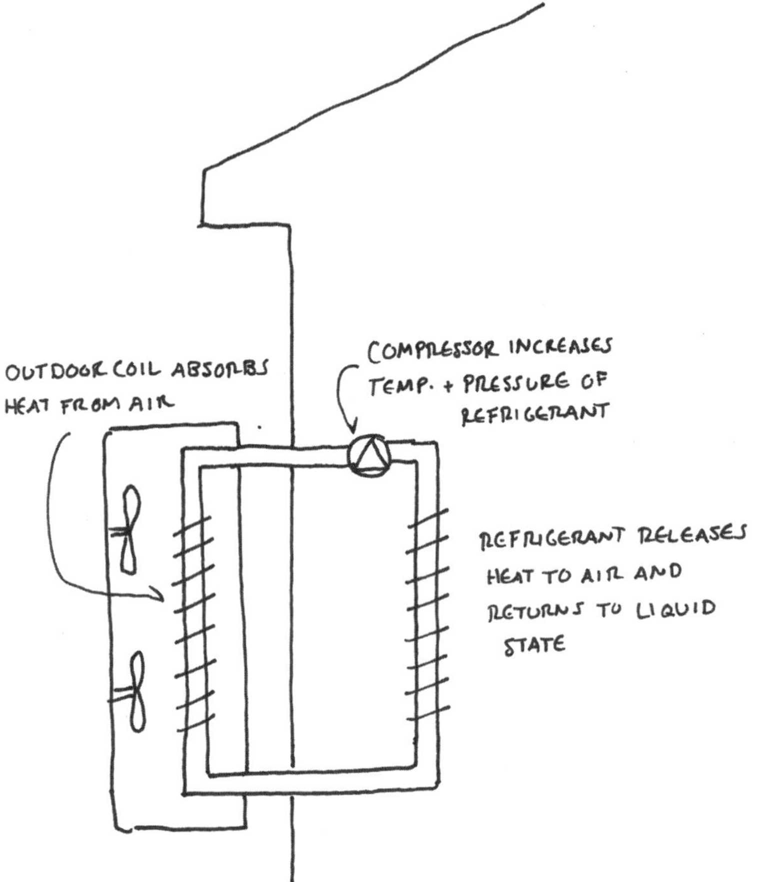

How the system works

A heat pump unit is mounted outside the home, and includes one or more large fans to move outdoor air over the heat exchanger coil to transfer air temperature to the refrigerant. The condenser/heat exchanger is contained in a plenum in the ductwork of the home, where the heat is delivered to the home via forced air or hydronics. The cycle is reversed to provide cooling in the summer.

A sub-set of ASHPs, called *ductless mini splits* operate on the same heat pump principle, with outdoor pumps absorbing heat from the ambient air. However, instead of supplying heat to a plenum in a central forced air system, they send the heated refrigerant to one or more (up to six, for some models) independent condenser "heads" or cassettes placed in a room, and room air is blown over the heat exchanger. This option eliminates the need for central ductwork and can provide well-distributed heat in an energy-efficient home for a lower cost.

System Output

ASHPs and ductless mini splits can be purchased with output ranging from 0.75–3 tons.

Boilers

Heat Production and Delivery

- Biomass heat production (wood and pellet fired)
- Fossil fuel heat production (gas and oil fired)
- Biofuel heat production (biodiesel and vegetable oil fired)
- Electric resistance heat production
- Hydronic heat delivery (can supply an air delivery system)
- Can be used for domestic hot water heating

System Components

- Combustion chamber (or resistance elements for electric units)
- Exhaust chimney (not required for electric units)
- Heat exchanger
- Direct vent (sealed) intake and exhaust (for combustion units)
- Circulation pump
- Temperature and pressure relief valve, expansion tank, air bleeder as required
- Controls as required

How the system works

When heat is required, a burner ignites. A water jacket heat exchanger places water in the path of the heat in a configuration for optimal heat absorption. When the water has reached the desired temperature, a circulating pump moves water through the system, providing a constant flow of heated output water.

Direct Heat — The heated water from the boiler is provided directly to the heating system. The boiler cycles on every time there is a call for heat. This system is common for space heating systems.

Indirect Heat — The heated water from the boiler is supplied to one or more storage tanks with a capacity based on expected heat requirements, and the storage tank is maintained at a set temperature by the boiler. When heat is called for, it is drawn from this tank rather than directly from the boiler. This system is common if the boiler is supplying both the space heating and domestic hot water systems, supplying multiple units, or if it is for domestic hot water alone. These systems do not require the boiler to cycle on each time there is a call for heat.

Condensing Boiler — Many newer combustion units (both direct and indirect) are condensing boilers. Exhaust gasses carrying waste heat are used to warm incoming water, putting more of the heat generated to use. The incoming pipes are carrying cooler water, and condensation can form on the pipes in the presence of the hot exhaust gasses. Condensate is removed through a drain. This type of boiler has higher efficiency than non-condensing models.

System Output

Residential boilers have outputs that range from 50,000–300,000 Btu/hr. Efficiency rates for condensing boilers range from 90–98%.

In most cases, the expressed efficiency is not the "true" thermal efficiency, which is often measured at an ideal steady-state and does not reflect real-world performance. The accepted standard of efficiency is *annual fuel utilization efficiency* (AFUE). The method for determining the AFUE for residential furnaces is the subject of ASHRAE Standard 103. Ensure that rated output for comparison between units is the AFUE.

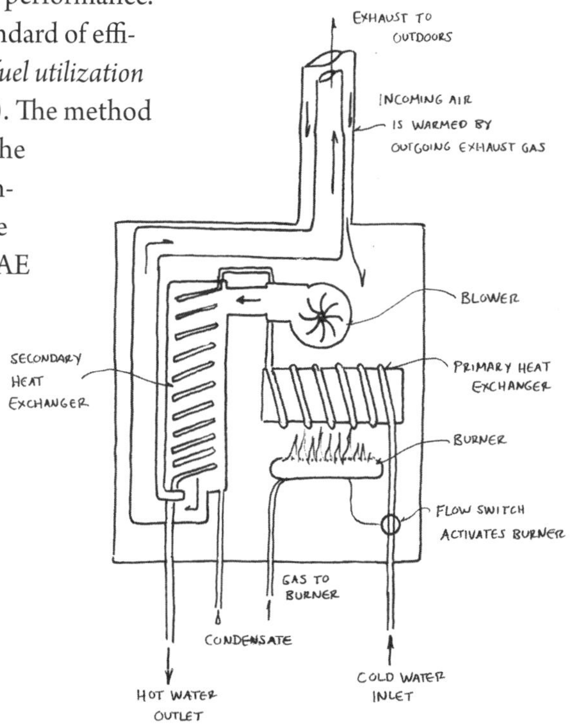

Tankless or on-demand heaters

Heat Production and Delivery

- Fossil fuel heat production (gas and oil fired)
- Electric resistance heat production
- Hydronic heat delivery
- Can be used for domestic hot water heating

System Components

- Combustion chamber (or resistance elements for electric units)
- Exhaust chimney (not required for electric units)
- Heat exchanger
- Direct vent (sealed) intake and exhaust (for combustion units)
- Circulation pump
- Temperature and pressure relief valve, expansion tank, air bleeder, and balance of system as required
- Controls as required

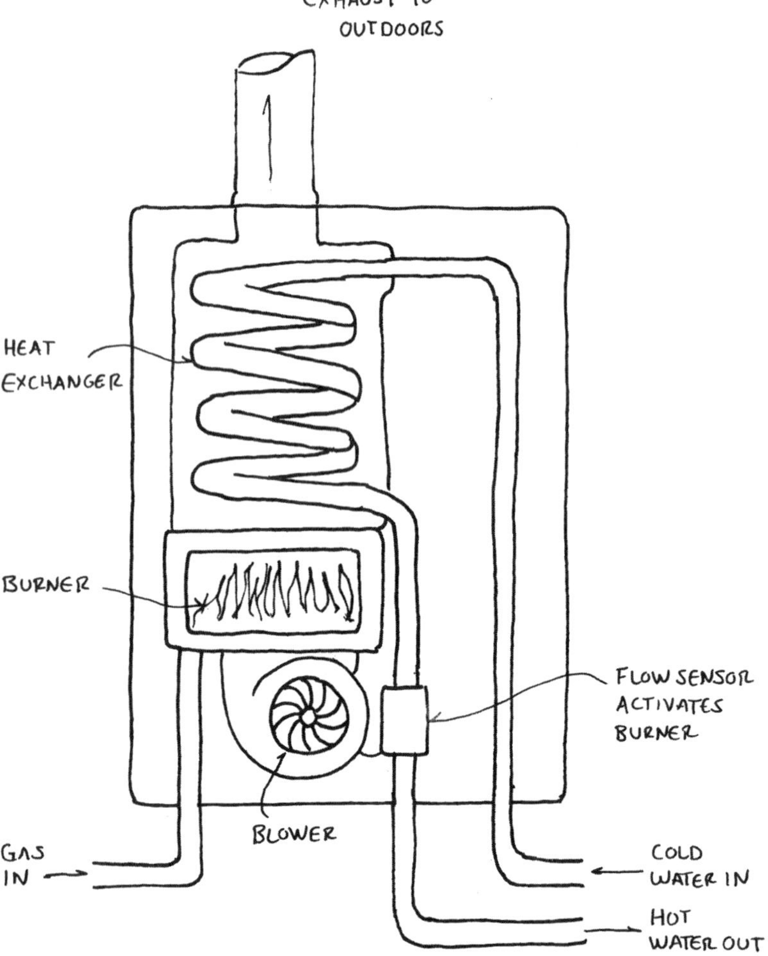

How the system works

Tankless or on-demand heaters are very similar to direct heat boilers. The difference is in the design of the heat exchanger, which is able to quickly impart a lot of heat to a relatively small amount of water. This can provide benefits for the production of domestic hot water, but is of dubious value for space heating systems.

A flow detector triggers the heater when water begins to move in the supply pipe. Advanced models also sense the temperature of incoming water and modulate the amount of heat accordingly.

System Output

Residential tankless heaters have outputs that range from 10,000–300,000 Btu/hr. Efficiency rates for gas boilers range from 90–98%. Electric resistance units are 100% efficient.

Ensure that rated output for comparison between combustion units is the AFUE.

Tank or batch heaters

Heat Production and Delivery

- Biomass heat production (pellet or wood fired)
- Fossil fuel heat production (gas and oil fired)
- Biofuel heat production (biodiesel and vegetable oil fired)
- Electric resistance heat production
- Hydronic heat delivery (can supply an air delivery system)
- Can be used for domestic hot water heating

System Components

- Metal or glass-lined metal tank with required inlet and outlet ports
- Tank insulation
- Integrated heat exchanger(s) if required
- Direct vent (sealed) intake and exhaust (for combustion units)
- Pressure and temperature relief valve, pumps, expansion tank, and balance of system as required
- Controls as required

How the system works

A tank of the desired size (20–300 gallons are common for residential purposes) holds water, which is heated from inside the tank. Combustion-fired units have a burner and exhaust inside the tank, and heat is absorbed into the water through contact with the chimney. Electric units have resistance heating elements inside the tank. Water is raised to and maintained at the desired temperature, and drawn from the tank as required.

Some units have integrated heat exchanger(s) to allow for combined use of space heating and domestic hot water and/or combining the tank with solar hydronic input.

System Output

Residential water heater tanks have outputs that range from 10,000–200,000 Btu/hr, depending on their volume capacity and type of burner or electric element. Efficiency rates for gas-powered tanks range from 90–98%. Electric resistance units are 100% efficient.

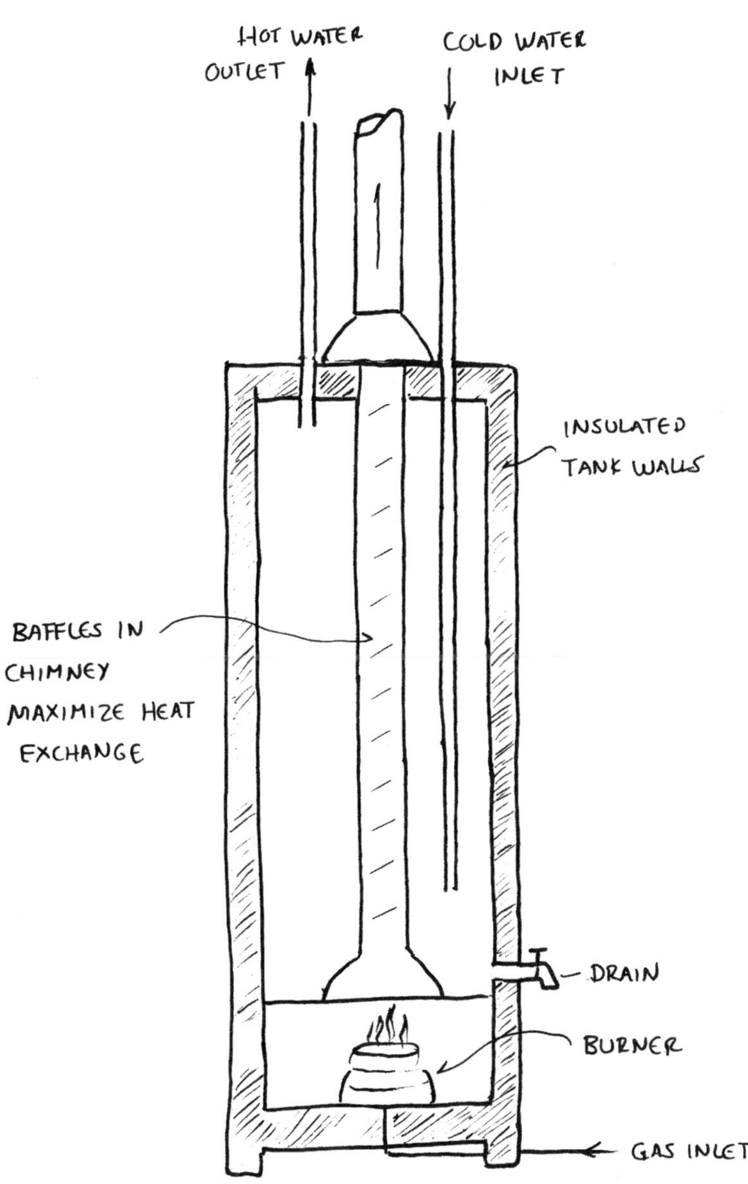

Forced air furnaces

Heat Production and Delivery

- Biomass heat production (wood and pellet fired)
- Fossil fuel heat production (gas and oil fired)
- Biofuel heat production (biodiesel and vegetable oil fired)
- Electric resistance heat production
- Forced air delivery

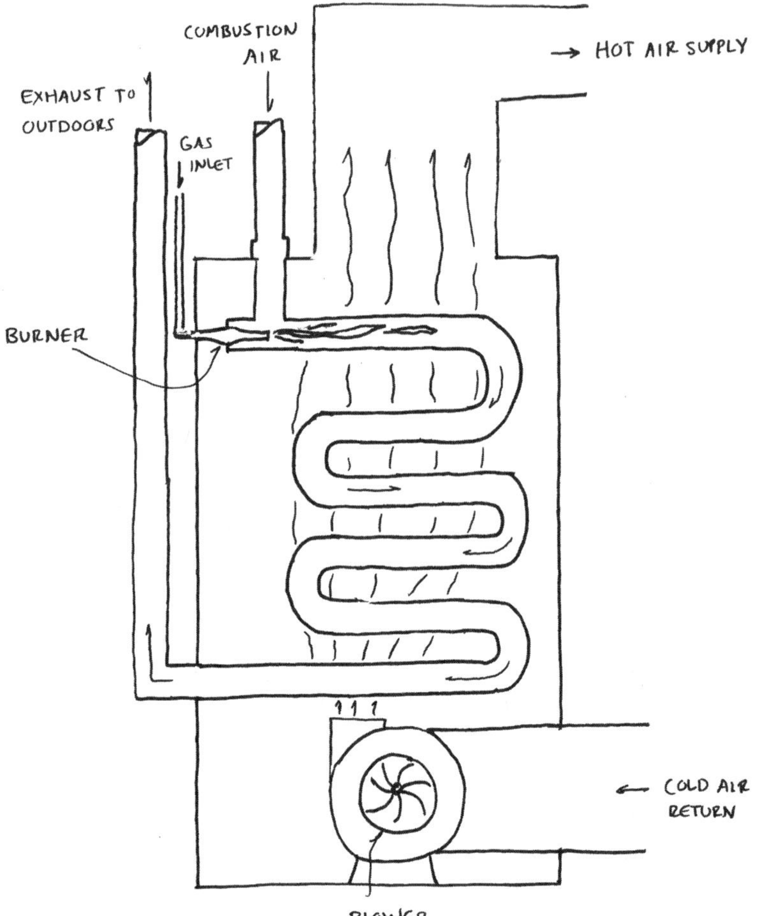

System Components

- Combustion chamber (or resistance elements for electric units)
- Heat exchanger and plenum
- Direct vent (sealed) intake and exhaust (for combustion units)
- Circulation fan
- Controls as required

How the system works

When heat is required, a burner ignites (or electric elements are powered). A heat exchanger plate absorbs heat from the flame/element and transfers it to the air being blown through the plenum.

The ductwork has both heat supply ducts and cold air return ducts. The cold air return ducts congregate at the plenum, where the air is reheated and forced through the supply ducts to outlets throughout the home. A filter is situated near the plenum to remove dust and allergens as the air is circulated.

System Output

Residential furnaces typically have outputs that range from 40,000–150,000 BtU/hr. Efficiency rates range for combustion furnaces range from 89–96%. Electric resistance units are 100% efficient.

Ensure that rated output for comparison between units is the AFUE.

Pellet and wood stoves

Heat Production and Delivery

- Biomass heat production (wood and pellet fired)
- Passive air/radiant delivery system
- Hydronic heat delivery available for some systems
- Forced air delivery available for some systems
- Can be used for domestic hot water heating

System Components

- Burn chamber and ash pan
- Air inlet and controls
- Exhaust outlet and controls
- Fuel delivery system (pellet stoves)
- Blower fan (some units)
- Water heating coil or tank (some units)

How the system works

A freestanding unit has a burn chamber where a fire is lit, heating a combination of masonry and metal around the chamber.

Combustion air is introduced through a manually or mechanically controlled inlet. Ideally, this combustion air is supplied through a sealed inlet directly from outdoors. This is nearly always the case with pellet burning units, but not as common with wood stoves.

Exhaust is vented through a chimney. This is accomplished passively in some units (mostly wood stoves), relying on the warm temperature of spent exhaust gas to provide an upward draft. In some units (mostly pellet stoves), a fan is used to draw combustion gasses out of the burn chamber and through the chimney.

The heat from the fire makes the metal surface of the unit hot, and the heat is transferred radiantly and also through significant convection loops formed as air is heated by the stove.

A fan may be incorporated to actively move heat from the stove into the room. Minimal ductwork is incorporated in some units. Rare models feature the ability to heat water within the stove. This can be a contribution to hydronic space heating and/or domestic hot water.

Fuel for wood stoves is manually input directly into the firebox through an opening door. Pellet stoves use an auger to automatically deliver fuel for combustion from a supply hopper. This hopper must be filled with pellets as required. Large hoppers can be used to ensure long periods of unattended operation.

System Output

Rated outputs for wood stoves are estimations, and actual output will be dependent upon combustion air supply, chimney draft, and the species and dryness of firewood. Typical residential models range from 50,000–200,000 Btu/hr. Burn efficiency is also difficult to accurately establish, but manufacturer claims range from 65–80%.

Pellet stoves have more accurate ratings, as pellets have relatively uniform density and moisture properties. Typical residential models range from 30,000–160,000 Btu/hr. Manufacturers claim burn efficiencies ranging from 70–90%.

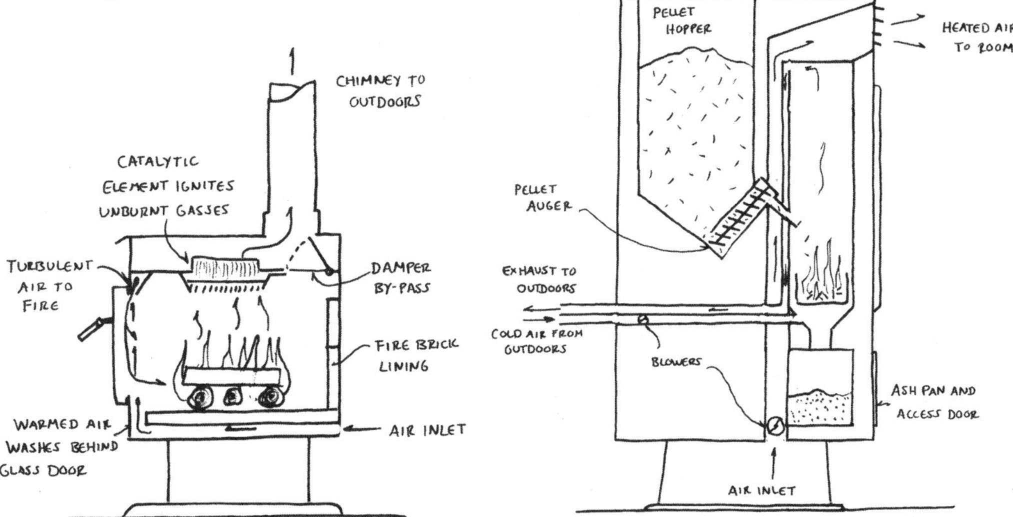

Masonry stoves and rocket mass heaters

Heat Production and Delivery

- Biomass heat production (wood fired)
- Passive air/radiant delivery system
- Hydronic heat delivery available for some systems
- Can be used for domestic hot water heating

System Components

- Burn chamber and ash pan
- Masonry chimney with non-linear pathway
- Masonry shroud (can be any kind of masonry, including clay)
- Air inlet and controls
- Cooking oven space (some units)
- Water heating coil or tank (some units)

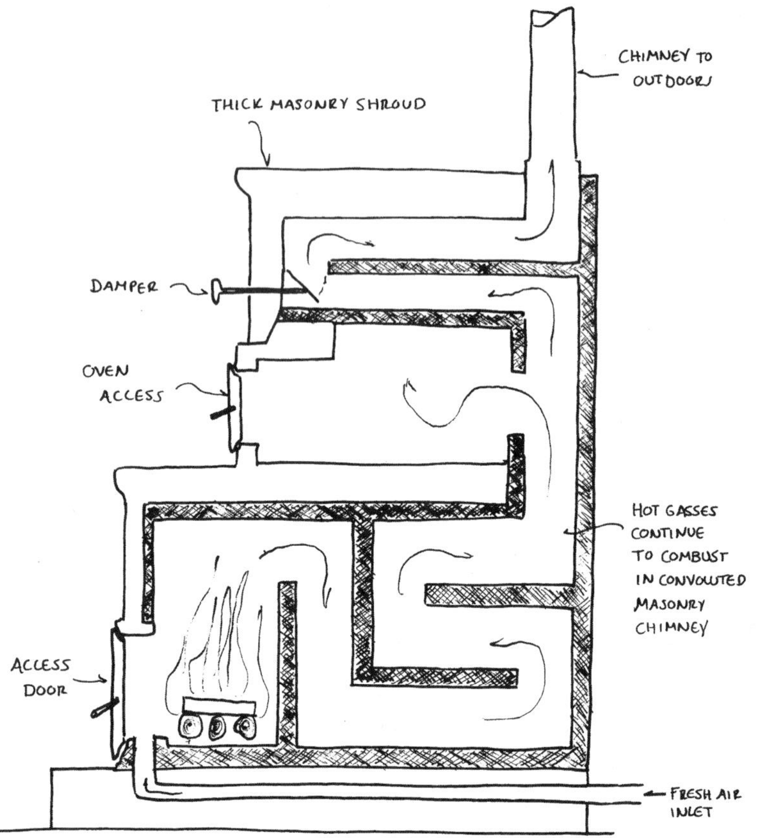

How the system works

A freestanding unit is placed, usually centrally, in the home. The unit has a burn chamber where a fire is lit.

Combustion air is introduced through a manually or mechanically controlled inlet. Ideally, this combustion air is supplied through a sealed inlet directly from outdoors. Masonry heaters do not use dampers on the air inlet, as the fire is burned at the highest possible temperature, requiring a significant column of air.

Exhaust gasses do not exit directly into a straight chimney, as with woodstoves. Instead, a masonry chimney with a non-linear path receives the gasses. Once the combustion chamber and chimney have become heated, the entire chimney pathway acts as a secondary burn chamber. At full temperature, practically 100% of the gasses are combusted within the chimney. These units are not limited in their burn temperatures by any metal components, and can burn at about 2,200–3,300°F (1200–1800°C).

Heat is absorbed into the masonry of the combustion chamber and chimney, and radiated to an outer sheathing of masonry, separated from the core by an air space. In this way, the interior of the home is not exposed to very high temperatures. Instead, the majority of the stored heat is slowly released to the room at comfortable temperatures. The outer sheathing can be made from a wide range of materials, from stone or brick to clay or tile. Spent gasses leave the building via a straight run of metal or masonry chimney.

System Output

There are no third-party ratings for masonry heaters, but estimates based on the volume of wood typically burned in an hour of operation range from 10,000–80,000 Btu/hr.

At proper operating temperature, combustion is 100% efficient. It will be less than that when starting from cold temperatures.

Electric resistance air heaters

Heat Production and Delivery

- Electric resistance

System Components

- Electric resistance element(s)

How the system works

There are two common forms of electric resistance heaters for residential use:

- **Baseboard heaters** — Electric resistance elements are encased in a metal shroud and mounted at the bottom of the wall. Air is heated by the elements and rises, circulating through the room.
- **Duct heaters** — Electric resistance elements are built into ductwork. Air is heated as the system fan moves air through the ducts.

System Output

Electric resistance heating elements come in a range of outputs, typically between 3,500 to 17,000 Btu/hr.

Ventilation systems

An integral part of the mechanical system for most new homes is active ventilation. Any home designed to achieve a high degree of energy efficiency is built airtight and will benefit from active ventilation to exhaust stale, humid, and/or contaminated air and replace it with fresh air from outside (see "Key Concepts for the Air Control Layer," in Chapter 6). Many code jurisdictions require active ventilation systems.

There are four kinds of home ventilation:

- **Exhaust only** — Extraction fans, typically in bathrooms and kitchen, are used to expel stale, humid air. This depressurizes the building, and make-up air is drawn through the building enclosure via leaks. This approach greatly increases the risks of pollution infiltration (humidity, mold spores, dust, radon, and other contaminants that can be pulled through cracks and leaks) and backdrafting of noxious gasses from combustion appliances.
- **Supply only** — Intake fans force outdoor air into the home. A flow regulator can control the amount of air entering the building. This system will pressurize the building, and air will be forced to leave the building through leaks. This approach greatly increases the risks of moisture damage to the building enclosure, especially in cold climates.
- **"Passive" ventilation** — Air movement for passive ventilation is driven by stack effect (a difference in temperature) and wind pressure. These factors can, when conditions are right, allow air to enter through open windows and/or leaks in the lower part of the building and exhaust through open windows and/or leaks in the upper part of the building. If windows are being used to encourage this ventilation, then the system is not truly passive, as it requires the homeowner to actively manage the opening and closing of the windows as

required. In hot or cold climates, this type of ventilation is not energy efficient. It makes sense to design passive ventilation strategies into your home as a resilience measure and for those times of year when it works effectively, but it shouldn't be the only ventilation strategy employed.

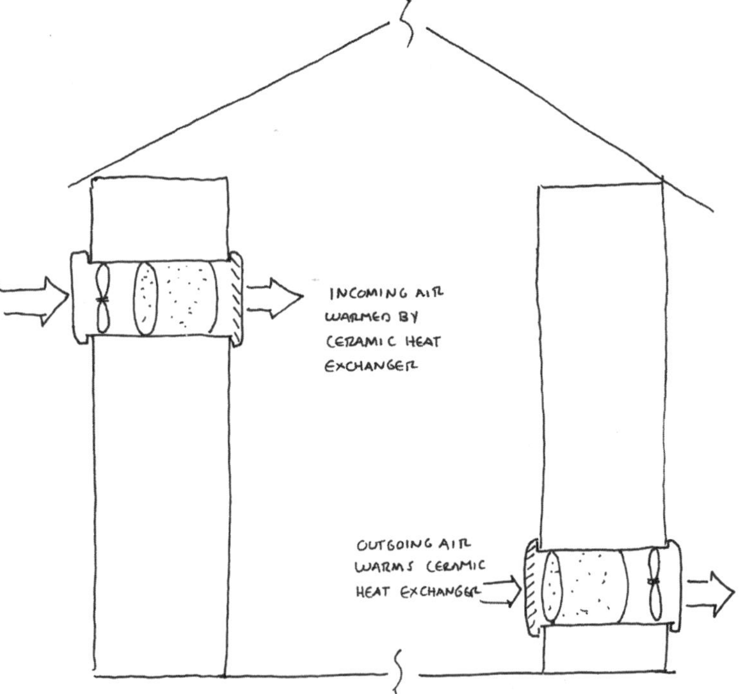

- **Balanced ventilation** — Heat recovery ventilation (HRV) and energy recovery ventilation (ERV) use the same system to manage exhaust and supply, increasing the comfort, efficiency, and health of the building. In an HRV system, the exhaust and intake air streams pass each other through an air-to-air heat exchanger, allowing the incoming air to be tempered by the outgoing air. In an ERV, the same heat exchange takes place, and in addition a humidity exchange also occurs, which can conserve additional energy, particularly when air conditioning is in use.

Balanced ventilation systems can be whole-house, ducted systems (with dedicated ductwork, or sharing ducts with a ducted heating/cooling system), or can be through-wall, ductless units.

Any balanced ventilation system offers the ability to filter supply air, to provide the highest quality of indoor air.

A balanced ventilation system is recommended for any home that is in a hot or cold climate, and where indoor air quality is a priority.

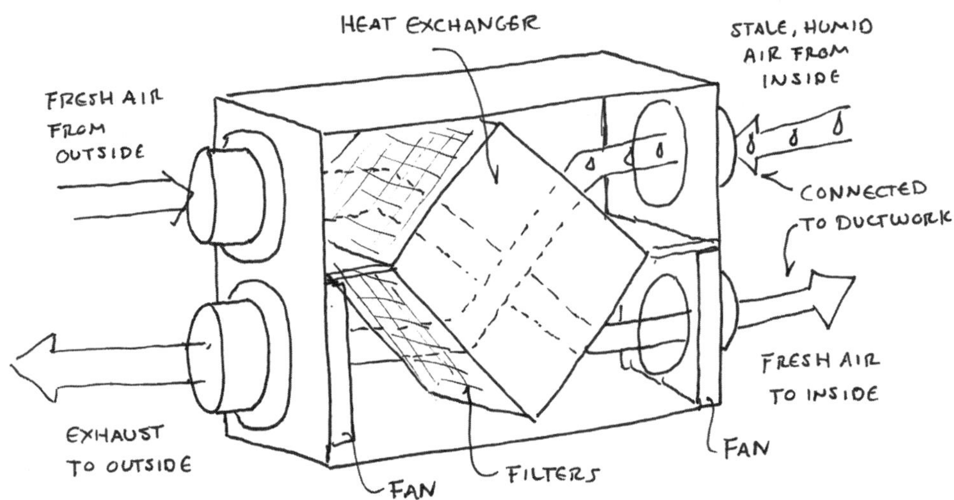

Chapter 12

Conclusion

MOST PEOPLE WANT TO LIVE IN A HOME that is comfortable and healthy while being light on the planet. Surprisingly, the criteria required to meet these seemingly basic goals are not central to the home design or construction industry, and to some degree you will need to take control to make sure your home meets your own criteria. In some cases, this will mean finding the right design and construction professionals, people who will understand your goals and be able to help you meet them. Or, it may mean taking control of a lot of the decisions yourself.

As this book has attempted to illustrate, it is vitally important to have clear goals for your project. Every choice you will face in the design and construction process becomes much easier if you have well-defined criteria to use for assessing your options.

This process can be engaging and fascinating. It can also be overwhelming. After digesting all the advice in this book (and beyond!), you will likely be striving to make the perfect choice at every junction, and you will inevitably come to the realization that you cannot. There may be competing criteria, budget limitations, code issues, availability concerns, and disagreements between all the partners in the project.

My best advice is to take some time to honestly rank your criteria in order of importance. If there are one or two goals you simply do not want to compromise, then be willing to make them your priorities and let some of the other criteria be of secondary importance. This often makes it easier to make those decisions that cannot satisfy every criterion you have for the project.

It can also help to focus on what I call "bulk achievement." Look at the decisions that will have the largest impact on your criteria — those that involve large quantities of material (insulation, cladding, flooring) and major mechanical systems. If you can meet your criteria in these areas, it's okay to compromise in some of the smaller areas.

The notion of "bulk achievement" can also refer to choices that come close to your goals, but don't quite meet them. For instance, if you want a home that is 90% more energy efficient than code, but budget restrictions limit you to 70% improvement, that is still a "bulk achievement." Or perhaps you can eliminate all questionable chemical content except for a few small components of the building; this is a huge improvement on the unregulated toxicity that is the norm.

By engaging consciously in a serious attempt to make a better building, you will undoubtedly create something that is far better than what would have been available to you in the conventional marketplace. Any improvements you can make in any of the criteria areas are significant and important.

We are, collectively, at the bottom edge of a learning curve to figure out how to maintain our current comfortable lifestyles while dramatically reducing our ecological impacts, our exposure to toxins, and our reliance on fossil fuels — all with an eye toward resiliency in a rapidly changing world. Your contributions to this collective learning will be important, so please share them. Publicize your successes, and don't be afraid to share those things that you could have done better. Show off your home, talk to relatives, friends, and neighbors about your motivations.

We have the ability to figure out how to design good sustainable homes, and you can play a part in the continual improvements that will help this notion to spread and take hold.

And, regardless of the goals you set and the ups and downs of the design and construction process, there are few satisfactions more profound than enjoying time with family and friends in a home that you have designed yourself!

Enjoy!

Endnotes

1. "Our Common Future: Report of the World Commission on Environment and Development," Oslo, March 1987, un-documents.net
2. PennState Extension report, "Water Quality: Septic System Failures," February 2012, extension.psu.edu
3. Wristen, Karen G. "The National Sewage Report Card (Number Two): Rating the Treatment Methods and Discharges of 21 Canadian Cities," Sierra Legal Defence Fund report, August, 1999.
4. Burton, Nancy Clark, and Chad Dowell. *Evaluation of Exposures Associated with Cleaning and Maintaining Composting Toilets.* Health Hazard Evaluation Report HETA 2009-0100-3135, U.S. Dept. Health and Human Services Centers for Disease Control and Prevention, July 2011.
5. Benhelal, E., et al. (2013). "Global strategies and potentials to curb CO_2 emissions in cement industry," *Journal of Cleaner Production* 51:142–161.
6. U.S. Census Bureau News. "New Residential Construction in December 2015," Joint Release by U.S. Department of Housing and Urban Development and U.S. Department of Commerce, Washington, D.C. , January, 2016, census.gov
7. Passipedia: The Passive House Resource. "Passive House Buildings in Use," n.d., passipedia.org
8. Klepeis, Neil E., et al. "The National Human Activity Pattern Survey (NHAPS): A Resource for Assessing Exposure to Environmental Pollutants," Ernest Orlando Lawrence Berkeley National Laboratory, n.d., indoor.lbl.gov
9. U.S. EPA & U.S. Consumer Product Safety Commission. (1995). "The Inside Story: A Guide to Indoor Air Quality," EPA 402-K-93-007, epa.gov
10. Testimony of Erik D. Olson Director, Health Program Natural Resources Defense Council before the Senate Committee on Environment and Public Works Hearing entitled "The Federal Role in Keeping Water and Wastewater Infrastructure Affordable," April 7, 2016.
11. "Characterization of Building-Related Construction and Demolition Debris in the United States," prepared for the U.S. Environmental Protection Agency Municipal and Industrial Solid Waste Division Office of Solid Waste, Report No. EPA530-R-98-010, epa.gov
12. *RCRA in Focus: Construction, Demolition, Renovation.* US Environmental Protection Agency Solid Waste and Emergency Response (5305W), EPA-530-K-04-005, September 2004, epa.gov
13. Lewis, Tanya. "Why People Don't Learn from Natural Disasters," July 9, 2013, livescience.com
14. "New Privately Owned Housing Units Started in the United States by Purpose and Design," US Census Bureau Building Permits Survey, Table Q1, 2013.
15. "LEED Platinum Projects in Canada," Canada Green Building Council, leed.cagbc.org
16. Lstiburek, Joe. "Insulations, Sheathings and Vapor Retarders," Research Report 0412, buildingscience.com
17. Mass, Carol. "Greenhouse Gas and Energy Co-Benefits of Water Conservation," POLIS Research Report 09-01, March 2009.
18. "Solids Inventory Control for Wastewater Treatment Plant Optimization," Issue No. 1.0, Federation of Canadian Municipalities and National Research Council, March 2004.
19. "Municipal Wastewater Treatment Indicator," Government of Canada, ec.gc.ca
20. "Ground-Source Heat Pumps (Earth-Energy Systems)," Government of Canada, oee.nrcan.gc.ca

Resources for Further Research

Design

Alexander, Christopher. *A Pattern Language: Towns, Buildings, Construction*. Oxford University Press, 1977.

Allen, Edward. *How Buildings Work: The Natural Order of Architecture*. Oxford University Press, 1995.

Baker-Laporte, Paula and Robert Laporte. *The EcoNest Home: Designing and Building a Light Straw Clay House*. New Society Publishers, 2015.

Ching, Francis D.K. *Building Construction Illustrated*. Wiley, 2014.

Ching, Francis D.K. *Architecture: Form, Space and Order*. Wiley, 2014.

Ching, Francis D.K. and Steven Juroszek. *Design Drawing*. Wiley, 2010.

Chiras, Daniel. *The Natural House: A Complete Guide to Healthy, Energy-Efficient, Environmental Homes*. Chelsea Green, 2000.

Pearson, David. *New Natural House Book: Creating a Healthy, Harmonious and Ecologically Sound Home*. Simon and Schuster, 1998.

Racusin, Jacob Deva, and Ace McArleton. *The Natural Building Companion: A Comprehensive Guide to Integrative Design and Construction*. Chelsea Green, 2012.

Susanka, Sarah. *Home by Design: Transforming Your House Into Your Home*. Taunton Press, 2004.

Wing, Charlie. *The Visual Handbook of Building and Remodeling: A Comprehensive Guide to Choosing the Right Materials and Systems for Every Part of Your Home*. Taunton Press, 2009.

Foundation systems

AAC

Moore, Ellie. *Autoclaved Aerated Concrete (AAC): A Study on AAC for Low-Rise Structures in Southern California*. LAP Lambert, 2014.

Earthbag

Geiger, Owen. "Earthbag Building: Earthbag Building Guide." *Earthbag Building: Earthbag Building Guide*. N.p., n.d. Web. 13 Apr. 2013.

Hunter, Kaki, and Donald Kiffmeyer. *Earthbag Building: The Tools, Tricks and Techniques*. Gabriola Island, BC: New Society Publishers, 2004.

Khalili, Nader, and Iliona Outram. *Emergency Sandbag Shelter and Eco-village: Manual — How to Build Your Own with Superadobe/ Earthbag*. Hesperia, CA: Cal-Earth, 2008.

Khalili, Nader. *Ceramic Houses and Earth Architecture: How to Build Your Own*. Hesperia, CA: Cal-Earth, 1990.

Wojciechowska, Paulina. *Building with Earth: A Guide to Flexible-form Earthbag Construction*. White River Junction, VT: Chelsea Green, 2001.

CMU

The Complete Guide to Masonry and Stonework. Chanhassen, MN: Creative, 2006.

Concrete

Fine Homebuilding (COR). *Foundations and Concrete Work: Revised and Updated*. Ingram, 2012.

King, Bruce. *Making Better Concrete*. Green Building Press, 2005.

King, Bruce, Ed. *New Carbon Architecture*. New Society Publishers, 2017.

Rammed earth tires

Hewitt, Mischa, and Kevin Telfer. *Earthships in Europe*. Watford, UK: IHS BRE Press, 2012.

McConkey, Robert. *The Complete Guide to Building Affordable Earth-Sheltered Homes: Everything You Need to Know Explained Simply*. Ocala, FL: Atlantic, 2011.

Prinz, Rachel Preston. *Hacking the Earthship: In Search of an Earth Shelter That Works for Everybody*. Archinia, 2015.

Reynolds, Michael E. *Earthship: Engineering Evaluation of Rammed-Earth Tire Construction*. S.l.: S.n., 1993.

Reynolds, Michael E. *Earthship: Evolution beyond Economics.* Taos, NM: Solar Survival Architecture, 1993.

Reynolds, Michael E. *Earthship: How to Build Your Own.* Taos, NM: Solar Survival Architecture, 1990.

Reynolds, Michael E. *Earthship: Systems and Components.* Taos, NM: Solar Survival, 1991.

Reynolds, Michael. *Comfort in Any Climate.* Taos, NM: Solar Survival, 2000.

Screw piers

Perko, Howard A. *Helical Piles: A Practical Guide to Design and Installation.* Hoboken, NJ: Wiley, 2009.

Stone

Cramb, Ian. *The Art of the Stonemason.* White Hall, VA: Betterway Publications, 1992.

Flynn, Brenda. *The Complete Guide to Building with Rocks and Stone: Stonework Projects and Techniques Explained Simply.* Ocala, FL: Atlantic Group, 2011.

Gallagher, A. Robert, Joe Piazza, and Sean Malone. *Building Dry-Stack Stone Walls.* Atglen, PA: Schiffer, 2008.

Long, Charles K. *The Stonebuilder's Primer: A Step-by-Step Guide for Owner-Builders.* Willowdale, Ont.: Firefly, 1998.

McRaven, Charles. *Building Stone Walls.* Pownal, VT: Storey, 1999.

McRaven, Charles. *Building with Stone.* Pownal, VT: Storey Communications, 1989.

McRaven, Charles. *Stonework: Techniques and Projects.* Pownal, VT: Storey, 1997.

Walls

Stud framing

Burrows, John. *Canadian Wood-Frame House Construction.* Ottawa, ON: Canada Mortgage and Housing Corporation, 2006.

Newman, Morton. *Design and Construction of Wood-Framed Buildings.* McGraw-Hill, 1995.

Simpson, Scot. *Complete Book of Framing: An Illustrated Guide for Residential Construction.* Kingston, MA: RSMeans, 2007.

Thallon, Rob. *Graphic Guide to Frame Construction: Details for Builders and Designers.* Newtown, CT: Taunton, 2000.

Tollefson, Chris, Fred P. Gale, and David Haley. *Setting the Standard: Certification, Governance, and the Forest Stewardship Council.* Vancouver, BC: UBC Press, 2008.

Timber framing & post and beam

Roy, Robert L. *Timber Framing for the Rest of Us.* Gabriola Island, BC: New Society Publishers, 2004.

Bensen, Tedd. *The Timber-Frame Home.* London: Taunton, 1997.

Sobon, Jack, and Roger Schroeder. *Timber Frame Construction: All About Post and Beam Building.* Pownal, VT: Storey, 1984.

Stirling, Charles. *Timber Frame Construction: An Introduction.* Garston, UK: BREhop, 2004.

Chappell, Steve. *A Timber Framer's Workshop: Joinery, Design and Construction of Traditional Timber Frames.* Brownfield, ME: Fox Maple, 1998.

Bingham, Wayne J., and Jerod Pfeffer. *Natural Timber Frame Homes: Building with Wood, Stone, Clay, and Straw.* Salt Lake City, UT: Gibbs Smith, 2007.

Beaudry, Michael. *Crafting Frames of Timber.* Montville, ME: Mud Pond Hewing and Framing, 2009.

Law, Ben, and Lloyd Kahn. *Roundwood Timber Framing: Building Naturally Using Local Resources.* East Meon, UK: Permanent Publications, 2010.

SIPs

Magwood, Chris. *Essential Prefab Straw Bale Construction.* New Society Publishers, 2016.

Lstiburek, Joseph. *Builder's Guide to Structural Insulated Panels (SIPS) for All Climates.* Building Science Press, 2008.

Adobe block

Byrne, Michael, Dottie Larson, and Amy Haskell. *New Adobe Home.* Layton, UT: Gibbs Smith, 2009.

McHenry, Paul Graham, Jr. *Adobe: Build It Yourself.* 2nd Revised Edition. Tucson, AZ: University of Arizona, 1985.

McHenry, Paul Graham. *Adobe and Rammed Earth Buildings: Design and Construction.* Tucson, AZ: University of Arizona, 1989.

Sanchez, Laura, and Al Sanchez. *Adobe Houses for Today: Flexible Plans for Your Adobe Home.* Santa Fe, NM: Sunstone, 2001.

Schroder, Lisa, and Vince Ogletree. *Adobe Homes for All Climates: Simple, Affordable, and Earthquake-Resistant Natural Building Techniques.* White River Junction, VT: Chelsea Green, 2010.

Van Hall, Michael. *The Cheap-Ass Curmudgeon's Guide to Dirt: Hand Building with Adobe, Papercrete, Paper-Adobe and More.* Tucson, AZ: Cheap-Ass, 2009.

Compressed earth block

Galer, Titane, Hubert Guillaud, Thierry Joffroy, Claire Norton, and Oscar Salaza. *Compressed Earth Blocks.* A Publication of Deutsches Zentrum für Entwicklungstechnologien — GATE, a Division of the Deutsche Gesellschaft Für Technische Zusammenarbeit (GTZ) GmbH in Coordination with the Building Advisory Service and Information Network — BASIN. Braunschweig, Germany: Vieweg, 1995.

Jaquin, Paul, and Charles Augarde. *Earth Building: History, Science and Conservation.* Bracknell, UK: IHS BRE, 2012.

Keefe, Laurence. *Earth Building: Methods and Materials, Repair and Conservation.* London: Taylor & Francis, 2005.

Morton, Tom. *Earth Masonry: Design and Construction Guidelines.* Bracknell, UK: IHS BRE, 2008.

Rael, Ronald. *Earth Architecture.* New York: Princeton Architectural Press, 2009.

Cob

Bee, Becky. *The Cob Builders Handbook: You Can Hand-Sculpt Your Own Home.* Murphy, OR: Groundworks, 1997.

Evans, Ianto, Linda Smiley, and Michael Smith. *The Hand-Sculpted House: A Philosophical and Practical Guide to Building a Cob Cottage.* White River Junction, VT: Chelsea Green, 2002.

Schofield, Jane, and Jill Smallcombe. *Cob Buildings: A Practical Guide.* Crediton, Devon: Black Dog, 2004.

Weismann, Adam, and Katy Bryce. *Building with Cob: A Step-by-Step Guide.* Totnes, Devon: Green, 2006.

Cordwood

Flatau, Richard C. *Cordwood Construction: A Log End View.* Merrill, WI: Flatau, 1997.

Flatau, Richard C., and Alan Stankevitz. *Cordwood and the Code: A Building Permit Guide.* Merrill, WI: Cordwood Construction, 2005.

Roy, Robert L. *Cordwood Building: The State of the Art.* Gabriola Island, BC: New Society Publishers, 2003.

Rammed earth

Easton, David. *The Rammed Earth House.* White River Junction, VT: Chelsea Green, 1996.

Jaquin, Paul, and Charles Augarde. *Earth Building: History, Science and Conservation.* Bracknell, UK: IHS BRE, 2012.

Keefe, Laurence. *Earth Building: Methods and Materials, Repair and Conservation.* London: Taylor & Francis, 2005.

McHenry, Paul Graham. *Adobe and Rammed Earth Buildings: Design and Construction.* Tucson, AZ: University of Arizona, 1989.

Minke, Gernot. *Earth Construction Handbook: The Building Material Earth in Modern Architecture.* Southampton, UK: WIT, 2000.

Morton, Tom. *Earth Masonry: Design and Construction Guidelines.* Bracknell, UK: IHS BRE, 2008.

Rael, Ronald. *Earth Architecture.* New York: Princeton Architectural Press, 2009.

Walker, Peter. *Rammed Earth: Design and Construction Guidelines.* Watford, UK: BRE hop, 2005.

Straw bale

Corum, Nathaniel. *Building a Straw Bale House: The Red Feather Construction Handbook.* New York: Princeton Architectural Press, 2005.

King, Bruce. *Design of Straw Bale Buildings: The State of the Art.* San Rafael, CA: Green Building, 2006.

Lacinski, Paul, and Michel Bergeron. *Serious Straw Bale: A Home Construction Guide for All Climates.* White River Junction, VT: Chelsea Green, 2000.

Magwood, Chris, and Chris Walker. *Straw Bale Details: A Manual for Designers and Builders.* Gabriola Island, BC: New Society Publishers, 2001.

Magwood, Chris, Peter Mack, and Tina Therrien. *More Straw Bale Building: A Complete Guide to Designing and Building with Straw.* Gabriola Island, BC: New Society Publishers, 2005.

The Straw Bale Alternative Solutions Resource. Victoria, BC: ASRi, 2013.

Insulation

Bynum, Richard T. *Insulation Handbook.* McGraw-Hill, 2000.

Doleman, Lydia. *Essential Light Clay Straw Construction.* New Society Publishers, 2017.

Magwood, Chris. *Essential Hempcrete Construction.* New Society Publishers, 2016.

Siddiqui, Sarfraz. *A Handbook on Cellulose Insulation.* Krieger Publishing, 1989.

Stanwix, William, and Alex Sparrow. *The Hempcrete Book.* Cambridge, UK: Green books, 2014.

Cladding and Roofing

Fine Homebuilding, Ed. *Siding, Roofing and Trim.* Taunton Press, 2014.

Jenkins, Joseph. *The Slate Roof Bible: Understanding, Installing and Restoring the World's Finest Roof.* Chelsea Green, 2003.

Marshall, Chris. *The Complete Guide to Roofing & Siding.* Cool Springs Press, 2013.

Sanders, Marjorie, and Roger Angold. *Thatches and Thatching: A Handbook for Owners, Thatchers and Conservators.* Ramsbury, UK, 2012.

Snodgrass, Edmund and Linda McIntyre. *The Green Roof Manual: A Professional Guide to Design, Installation and Maintenance.* Timber Press, 2010.

Flooring

Black & Decker, ed. *The Complete Guide to Tile.* Cool Springs Press, 2015.

Crimmel, Sukita Reay, and James Thomson. *Earthen Floors: A Modern Approach to an Ancient Practice.* New Society Publishers, 2014.

Jeffries, Dennis. *The Flooring Handbook: The Complete Guide to Choosing and Installing Floors.* Firefly, 2004.

Peterson, Charles. *Wood Flooring: A Complete Guide to Layout, Installation and Finishing.* Taunton Press, 2010.

Finishes

Crews, Carole. *Clay Culture: Plasters, Paints and Preservation.* Gourmet Adobe, 2009.

Edwards, Lynn, and Julia Lawless. *The Natural Paint Book: A Complete Guide to Natural Paints, Recipes and Finishes.* Rodale, 2002.

Weismann, Adam, and Katy Bryce. *Using Natural Finishes: Lime- and Earth-Based Plasters, Renders and Paints.* Green, 2008.

Water and Wastewater

Banks, Suzy, and Richard Heinichen. *Rainwater Collection for the Mechanically Challenged.* Tank Town Publishing, 2006.

Burns, Max. *Country and Cottage Water Systems: A Complete Guide to On-Site Water and Sewage Systems.* Cottage Life, 2010.

Cipollina, Andrea et al. *Seawater Desalination: Conventional and Renewable Energy Processes.* Springer, 2009.

Del Porto, David, and Carol Steinfield. *The Composting Toilet System Book: A Practical Guide to Choosing, Planning and Maintaining Composting Toilet Systems.* Center for Ecological Pollution Prevention, 2007.

Ingram, Colin. *The Drinking Water Book: How to Eliminate Harmful Toxins from Your Water.* Celestial Arts, 2006.

Jenkins, Joseph. *The Humanure Handbook: A Guide to Composting Human Manure.* Joseph Jenkins, 2005.

Kinkade-Levario, Heather. *Design for Water: Rainwater Harvesting, Stormwater Catchment and Alternate Water Reuse.* New Society Publishers, 2007.

Krishna, J.H. *The Texas Manual on Rainwater Harvesting.* Texas Water Development Board, 2005.

Parten, Susan M. *Planning and Installing Sustainable Onsite Wastewater Systems: A Detailed Guide to Sustainable Decentralized Wastewater Systems.* McGraw-Hill, 2010.

Wing, Charles. *How Your House Works: A Visual Guide to Understanding and Maintaining Your Home.* RS Means, 2007.

Woodson, R. Dodge. *Water Wells & Septic Systems Handbook.* McGraw-Hill, 2003.

HVAC

Bainbridge, David, and Kenneth Haggard. *Passive Solar Architecture: Heating, Cooling, Ventilation, Daylighting and More Using Natural Flows.* Chelsea Green, 2011.

Chiras, Daniel. *The Solar House: Passive Heating and Cooling.* Chelsea Green, 2002.

Egg, Jay, and Brian Howard. *Geothermal HVAC: Green Heating and Cooling.* McGraw-Hill, 2011.

Jenkins, Dilwyn. *Wood Pellet Heating Systems: The Earthscan Expert Handbook on Planning, Design and Installation.* Earthscan, 2010.

Laughton, Chris. *Wood Heating: A Guide to Pellet, Chip and Log Heating Systems.* CAT Publications, 2012.

Lloyd, Donal Blaise, and Lawrence Muhammad. *Geo Power: Stay Warm, Keep Cool and Save Money.* PixyJack Press, 2015.

Matesz, Ken. *Masonry Heaters: Designing, Building and Living with a Piece of the Sun.* Chelsea Green, 2010.

Pahl, Greg. *Natural Home Heating: The Complete Guide to Renewable Energy Options.* Chelsea Green, 2003.

Ramlow, Bob, and Nusz, Benjamin. *Solar Water Heating: A Comprehensive Guide to Solar Water and Space Heating Systems.* New Society Publishers, 2006.

Stojanowski, John. *Residential Geothermal Systems: Heating and Cooling Using the Ground Below.* Pangea Publications, 2010.

Thomas, Dirk. *The Woodburner's Companion: Practical Ways of Heating with Wood.* Alan C. Hood, 2006.

Wing, Charles. *How Your House Works: A Visual Guide to Understanding and Maintaining Your Home.* RS Means, 2007.

Wisener, Erica, and Ernie Wisener. *The Rocket Mass Heater Builder's Guide: Complete Step-by-Step Construction, Maintenance and Troubleshooting.* New Society Publishers, 2016.

Woodson, Roger Dodge. *Radiant Floor Heating.* McGraw-Hill, 2010.

Electrical Generation

Boxwell, Michael. *Solar Electricity Handbook: A Simple, Practical Guide to Solar Energy — Designing and Installing Photovoltaic Solar Electric Systems.* Greenstream, 2016.

Chiras, Daniel. *Power from the Sun: A Practical Guide to Solar Electricity.* 2nd ed. New Society Publishers, 2016.

Chiras, Daniel et al. *Wind Power Basics.* New Society Publishers, 2010.

Davis, Scott. *Microhydro: Clean Power from Water.* New Society Publishers, 2003.

Index

Page numbers in *italics* indicate tables.

About the Author

Chris Magwood is obsessed with making the best, most energy efficient, beautiful and inspiring buildings without wrecking the whole darn planet in the attempt.

Chris is currently the executive director of The Endeavour Centre, a not-for-profit sustainable building school in Peterborough, Ontario. The school runs three full-time, certificate programs: Sustainable New Construction, Sustainable Renovations, and Sustainable Design, and it hosts many hands-on workshops annually.

Chris has authored numerous books on sustainable building, including *Making Better Buildings* (2014), *More Straw Bale Building* (2005) and *Straw Bale Details* (2003). He is co-editor of the Sustainable Building Essentials series, and is a past editor of *The Last Straw Journal*, an international quarterly of straw bale and natural building. He has contributed articles to numerous publications on topics related to sustainable building and maintains a blog entitled "Thoughts on Building."

In 1998 he co-founded Camel's Back Construction, and over eight years helped to design and/or build more than 30 homes and commercial buildings, mostly with straw bales and often with renewable energy systems.

Chris is an active speaker and workshop instructor in Canada and internationally.

A Note About the Publisher

NEW SOCIETY PUBLISHERS is an activist, solutions-oriented publisher focused on publishing books for a world of change. Our books offer tips, tools, and insights from leading experts in sustainable building, homesteading, climate change, environment, conscientious commerce, renewable energy, and more — positive solutions for troubled times.

We're proud to hold to the highest environmental and social standards of any publisher in North America. This is why some of our books might cost a little more. We think it's worth it!

- We print all our books in North America, never overseas
- All our books are printed on **100% post-consumer recycled paper**, processed chlorine free, with low-VOC vegetable-based inks (since 2002)
- Our corporate structure is an innovative employee shareholder agreement, so we're one-third employee-owned (since 2015)
- We're carbon-neutral (since 2006)
- We're certified as a B Corporation (since 2016)

At New Society Publishers, we care deeply about *what* we publish — but also about how we do business.

New Society Publishers

ENVIRONMENTAL BENEFITS STATEMENT

For every 5,000 books printed, New Society saves the following resources:[1]

49	Trees
4,465	Pounds of Solid Waste
4,913	Gallons of Water
6,408	Kilowatt Hours of Electricity
8,116	Pounds of Greenhouse Gases
35	Pounds of HAPs, VOCs, and AOX Combined
12	Cubic Yards of Landfill Space

[1]Environmental benefits are calculated based on research done by the Environmental Defense Fund and other members of the Paper Task Force who study the environmental impacts of the paper industry.